PLUMB IT

Bob Tattersall

CONTENTS

COLLINS

Introduction

There was a time when you had to warn the do-it-yourselfer not to tackle plumbing, apart from simple jobs such as fitting a new washer on a tap; but all that has changed. Several developments are responsible: pipes are now made of easy-to-work materials. Also, new devices have made it easy to join the pipes and connect them up – to taps and cisterns, for instance.

As a result, the amateur plumber can take on a whole range of improvements and installations that would previously have been thought right out of the question.

This is not a manual on how to install a complete plumbing system in an empty house – a task few do-it-yourselfers would ever be called upon to face. But the book does show you how to care for your existing plumbing – how to maintain it, carry out repairs, and extend it to add extra facilities that will give you and your family a better home – and a better life.

Very small cloakrooms are often the rule in modern houses. Stick to a small basin and don't attempt to hide the plumbing.

Far left *If you are building a new bathroom, or replacing old fittings, why not consider plumbing in two basins?*

Above left *It's a good idea to try to keep all your plumbing in one part of the kitchen. Fitting the dishwasher next to the sink makes for minimal pipework. Ideally it should be on an outside wall, close to the drain.*

Left *This large nursery room has its own washing machine and sink unit concealed behind folding louvred doors. This seems a very practical solution to the mounds of washing produced by small children, but do keep an eye on the plumbing. A flood on the first floor could prove both inconvenient and expensive.*

HOW A PLUMBING SYSTEM WORKS

Before you can do any jobs on your own plumbing system, you need to understand it. There are, in fact, two aspects to it. You have to bring water to where you want it – the supply; and get rid of it when you're finished with it – the waste. Let us look at both, initially, in terms of a conventional small house built during the last 50 years or so.

Supply

First, the supply. Water enters your property (not necessarily your house) underground, via a branch pipe from the Water Board's main, running under the street. From the point at which it does so, it becomes your responsibility. Near the garden gate or in the street, there should be a *stop tap* to allow you to turn off the supply in an emergency or when work needs to be done. The pipe then continues underground and emerges in the house, usually near the kitchen sink. There should be a stop tap here, too. Just beyond the stop tap, there will be a branch line to the sink's cold tap, so that water used for cooking and drinking will be pure from the main.

The main pipe now continues upwards and is known as the *rising main*. It feeds a tank called the *cold storage tank*, which is usually in the loft (it's the one that freezes up in winter if it isn't lagged properly), but it could be in some other part of the house. Indeed, it will have to be if the house has a flat roof (and therefore no loft).

The Cold Storage Tank

This tank usually supplies all the cold outlets (taps) in the house, except for the one at the kitchen sink. It also feeds with cold water any water heating system powered by a boiler. However, it could be that, in order to save on pipework, a builder has connected some, or even all, cold outlets direct to the rising main. Many water authorities frown on this.

Water rises up to the cold storage tank because of the pressure of the mains. It is then fed to the various outlets by gravity – which is why the tank must be sited high in the house. Water stored in this tank might get contaminated by dust, insects, birds etc, so water from cold taps not fed directly by the mains should not be drunk without first being boiled.

At various points throughout the system there will be other stop taps – the cold feed to a wc, for instance, or to the hot water system – so that these can be isolated when you need to work on them.

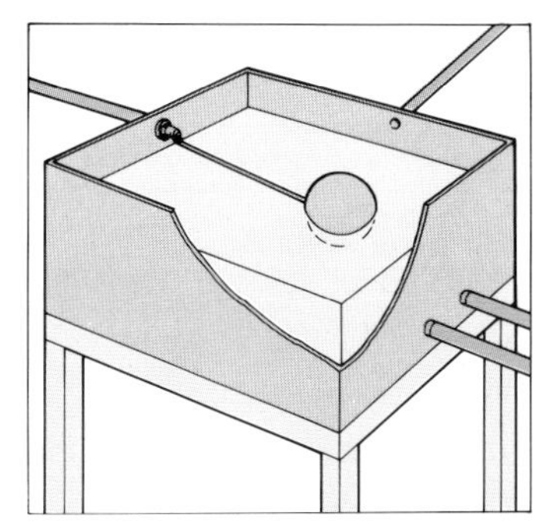

The entry of water into a cistern – on a wc or the cold storage tank – is controlled by what is known as a *ball valve*. The valve has an arm, on the end of which is a metal or plastic ball that floats on the surface of the water. When the level in the tank is low, the ball drops, taking the arm with it, and that opens the valve. As more water is admitted, the level rises, raising the ball and arm, and so the valve is shut off. In case anything should go wrong and too much water comes in, there is an overflow pipe so that the excess water will be carried away safely outside, instead of flooding the house. Such an overflow (the term is used to describe the flow of excess water, as well as the pipe itself) also draws your attention to the fault.

Boilers

The most common form of domestic water heating is by means of a boiler – free-standing, or a back boiler that is part of a fireplace. (Plumbers use the word 'domestic' in this context not as the opposite of 'industrial', but to refer to the water drawn off the household taps, as opposed to the water in the radiators of a central heating system.) The boiler may be fired by gas, solid fuel or oil. The heated water is stored, ready for use, in a tank known as the hot cylinder, that is nearly always located in an airing cupboard.

The system is usually a gravity one, and functions because hot water is lighter (in weight) than cold. Thus, in any container, water that is heated will rise to the top, and cold water will drop to the bottom. How this works in a water heating system is that the heated water rises from the boiler to the hot cylinder by means of a pipe known as the *flow*, and cooler water is pushed down (by gravity) to the boiler via the *return pipe*. Thus there is

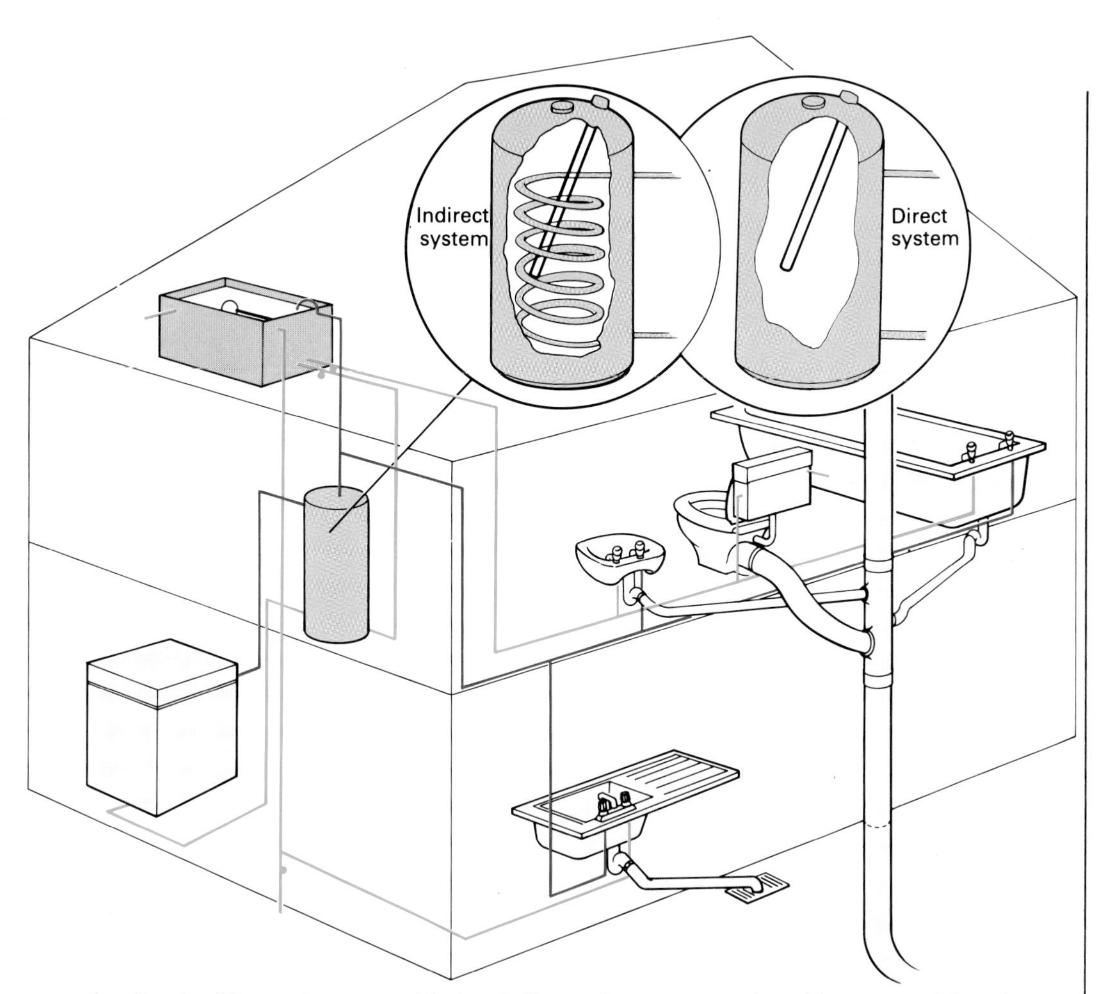

a continual cycle of flow and return whenever heat is supplied by the boiler. The flow and return pipes are collectively called *primaries*

For a gravity system to work properly, the hot cylinder must be placed higher up than the boiler. Where this isn't possible – in one-level homes such as flats, for instance – the water has to be forced from boiler to cylinder artificially, and so you get *pumped primaries*.

The hot cylinder may well have an electric immersion heater. At one time, these were fitted as the sole means of water heating, but are nowadays more likely to be intended to supplement a boiler during breakdowns and in summer.

Modern boilers and immersion heaters are controlled by thermostats to stop them from overheating. However, just in case things do get out of hand, a pipe (called the *vent pipe*) rises up from the top of the hot cylinder to the cold tank, over which it is bent like a shepherd's crook, so that any overheated water is discharged into the tank. If, as a result, the cold tank becomes too full, excess water will discharge via the overflow.

The draw-off point (the pipe that carries hot water to the taps etc) is in the vent pipe, just above the cylinder. This is where the water will be hottest. Cold water to replace the hot that is drawn off is normally fed from the cold storage tank into the bottom of the cylinder.

A drain cock will be fitted at the lowest point of the system, usually near the boiler, so that the water can be drained off when necessary.

The boiler system may be direct or indirect. In a *direct system*, the water that is heated in the boiler goes directly to the hot cylinder to be drawn off at the taps. With an *indirect system*, however, it is fed to a small inner cylinder or coil inside the hot cylinder, which thus becomes hot. It is the heat from the coil that warms up the water in the outer cylinder, which will then go to the taps. This way, the water in the boiler and the water you actually use are quite separate. Indirect cylinders are usually confined to homes with central heating, so that radiator and domestic will never get mixed up. However, you do find indirect cylinders in a system that merely heats domestic water.

If you live in an older house (where a modern plumbing system is a later addition), in a house with a rambling complicated ground plan, or in a flat (especially a conversion in an old house), the layout of the supply may not be as neat as that shown in the diagram, but it will be based on the general principles outlined.

Try to get to know your plumbing system. Go round your house, following the pipes, with the book in hand open at the previous page, until you can understand its layout. In particular, try to determine which fittings and sections are controlled by various stop taps. Shut off a stop tap and then see which taps on baths, sinks, basins, lavatories etc will then not work. Beware of drawing off too much water from a hot system with the tap shut off and the boiler working. You should turn the boiler off.

Instantaneous Heaters

In small flats created by converting large houses, and in older homes built without a water heating system, but later modernised on a budget, it may be too complicated to add the sort of system described above. Water may then be heated by a gas or electric instantaneous heater. There is no hot cylinder – hot water is not stored – and thus, no airing cupboard.

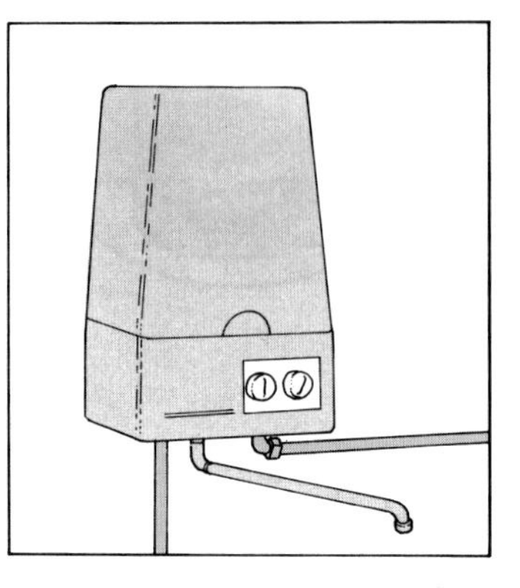

The heater will be connected directly to the mains, and its gas jets or electric element will come into play only when a tap is turned on, and water flows. Such heaters can be single or multi-point. A single point will be placed over one sink or bath and provide hot water just at that spot. However, pipes will carry water from a multi-point to more than one tap, usually to all those in the home.

Single point instantaneous heaters are sometimes fitted to a house with a full-scale boiler storage system as a means of providing hot water during the summer.

The Head of Water

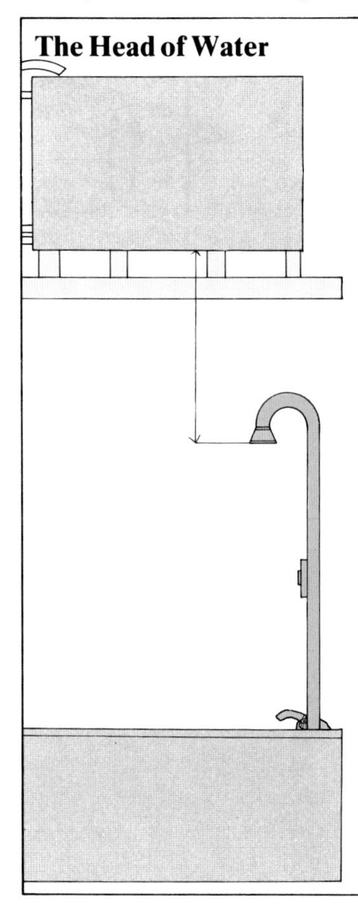

The difference in height between the cold storage tank and the outlet at which the water emerges – tap or wc valve, for instance – is known as the *head of water*. Plumbers usually take the measurement from the bottom of the tank, for the water level in it will often be low. The head of water determines the pressure of water at the outlet, and thus the force with which it will come out. A good head of water is always necessary to give a strong flow – you couldn't, for instance, have a tap higher than the cold tank – but it is especially important when you install a shower. Note that the pressure at the hot taps, just as surely as the cold, is determined by the head of water between outlet and cold tank. The hot cylinder plays no part in determining the pressure; from this point of view it is merely a swollen-up section of pipe through which the water passes on its way from cold tank to tap.

Waste

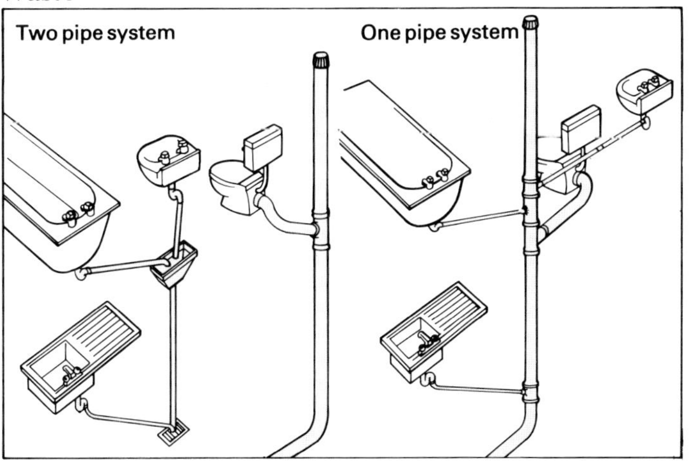

Two types of domestic drainage system are used in British homes: the two-pipe system, and the one-pipe system (sometimes called the single-stack system). Most homes have the former. In this, the waste from the wc is kept apart from the waste from sinks, baths, washbasins, bidets etc (known collectively as 'fittings'). The wc discharges into a soil pipe that connects to the local authority sewer. The soil pipe also extends upwards – well away from windows – to act as a vent for the escape of sewer gas. It is topped by a wire cage to stop rubbish getting inside it and causing blockage.

Gullies

Waste pipes from ground floor fittings discharge into open gullies that are connected to the drains. Baths, basins and other fittings on upper storeys empty into a type of open gulley, known as a *hopper*, fitted to the top of a downpipe. This pipe, in turn, conveys the water to a ground floor open gulley – sometimes the same one into which the ground floor waste pipes discharge.

Modern and high-rise homes may have one-pipe systems. In this, both wcs and fittings discharge into one pipe (usually inside the building) that carries all the waste to the drains. This pipe (or stack) was at onc time made of cast iron, but nowadays only plastic is used.

It is easy to tell which type of drainage system your home has.

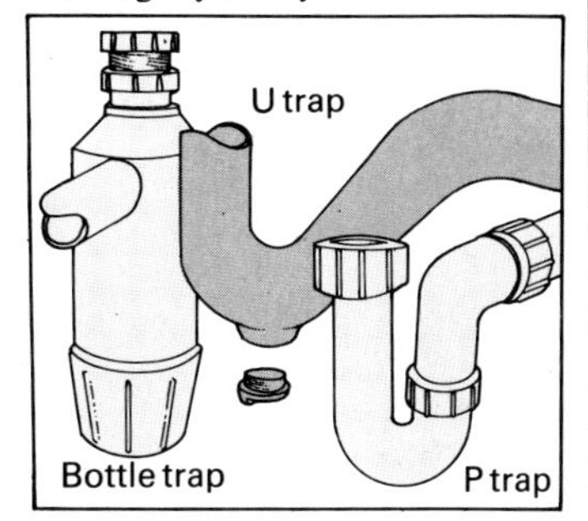

The Trap

One feature of all waste pipes is that they have a trap. This is a device designed to trap a small reservoir of water to stop drain smells from coming up the pipe and into the home. A wc trap is fitted at the back of (or sometimes forms part of) the pan. The trap for sinks and other fittings is a separate item inserted into the waste pipe during installation.

The Rules

Any changes you make to your home's plumbing system must be carried out in accordance with your local water authority's regulations. These will cover the sizes of various tanks and cisterns, plus precautions to stop appliances that mix hot and cold water (eg showers and mixer taps) from sucking water that has been stored in a tank (and therefore won't be fit for drinking) back into the mains, and so on.

You don't have to get permission to carry out plumbing work, as you do with the planning laws or building regulations, but the water authority has the right to inspect your home and to demand alterations.

The rules change from time to time, and vary from one authority to another. You can carry out minor repairs (fitting new washers, stopping overflows, curing leaks etc) without consultation, but you should always check before doing anything major.

Building regulation approval is needed if you want to alter or extend the waste water system. Consult the Building Control Department of your local authority.

Kitchens

A few plumbing improvements can make your kitchen a much easier place in which to work. For instance, you can fit new, better taps in place of the old, change two taps for a mixer, and so on.

Most important of all, you should make sure that *washing machines* and *dishwashers* are properly plumbed in. You will find it much more convenient than constantly having to wheel the machine out, attach a hose to the taps, and clip a waste to the edge of the sink (with the risk of the pipe coming adrift and causing a flood).

Above right *A modern stainless steel sink unit has been set into a tiled work surface. The mixer tap unit swings to reach both sinks, and there is a spray attachment and waste disposal.*

Above *An old pine cupboard has been converted to hold a brightly coloured, square sink. Taps have been matched to the sink and mounted behind.*

Opposite *The space under the stairs outside this basement kitchen has been used as a laundry area. Maybe you too have some unused space like this, that could get the washing out of the kitchen.*

INSTALLING A SUPPLY

Whenever you want to install a new fitting, or plumb in an electrical appliance, you have to lay on a supply of water to it. At one time water supply pipes were made of lead (the word plumb comes from the Latin word for lead) and between the wars steel pipes were used. Modern plumbing, however, uses copper pipe and, increasingly, plastic.

If your home has lead or steel supply pipes, the system is obsolete and needs complete renewal. You might even qualify for a home improvement grant. Copper pipe is the most usual, and it is easily worked by the do-it-yourselfer.

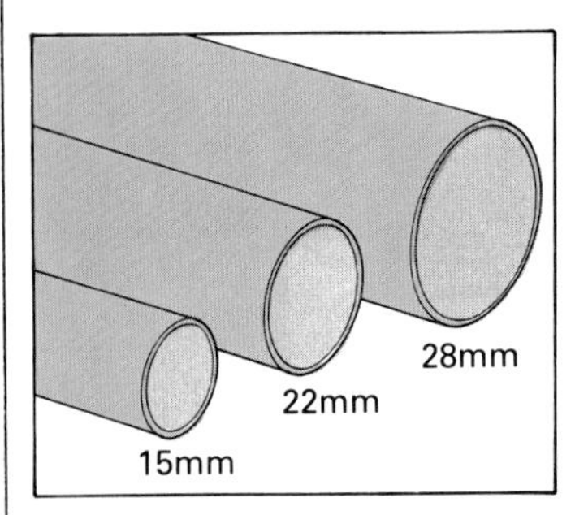

Domestic copper supply pipe is sold in metric sizes, and there are three main diameters – 28mm for the primaries, 22mm for main runs, and 15mm for *branch lines*. If your existing installation is 15 or more years old, the pipes will be in the imperial sizes of 1 inch, ¾ inch and ½ inch. These, of course, are not direct translations of the metric, but this doesn't matter. The metric figures refer to the external diameter of the pipe, while the imperial ones refer to the internal diameter. The 28 and 15mm sizes are directly interchangeable with the 1 and ½ inch sizes, and when you wish to connect 22mm to ¾ inch pipe, you must use an adaptor.

Working Copper Tube

There are four different operations to be carried out on plumbing pipework. When you want to install a water supply to a fitting, all, or most of them, will be necessary.

The pipe must be cut to length; it will need to be joined (eg to make two short lengths long enough, or to take a branch line at an angle from a straight run); it may have to be bent to go around corners; and it will need to be connected to fittings, such as taps or a cylinder.

Cutting

As copper is a soft metal, the supply pipes can easily be cut to length with a hacksaw. However, there is a tendency for the cut end to flatten slightly, and this makes it difficult to make a waterproof joint. Also, the cut end must be truly square, and it is not easy to ensure this with a hacksaw.

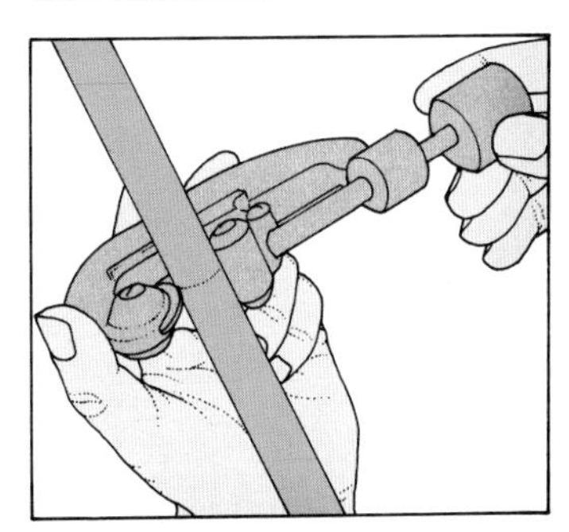

It is better to use a proper *pipe cutter* that will not flatten the end, and will make a cut you can be sure is square. Small portable cutters, intended for the do-it-yourselfer, are not very expensive.

However, when you are cutting into pipework that is already installed and fixed close to a wall, there is often not enough room to wield a pipe cutter. Then you must use a hacksaw.

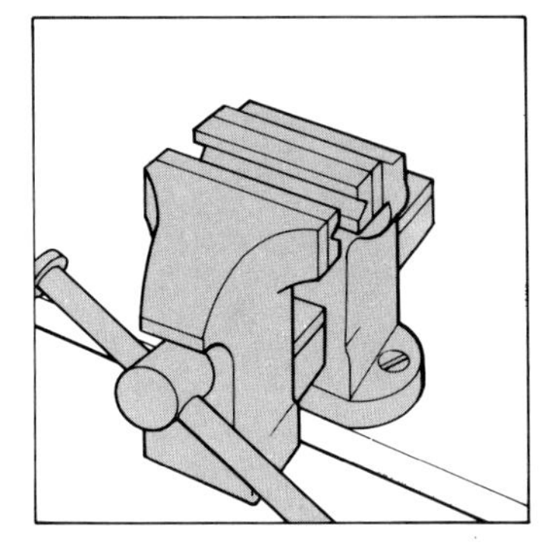

Never hold pipe in a woodworker's vice: use a proper pipe vice (or a Workmate) instead. This will prevent the pipe being dented and impeding the flow of water in use. Always aim to avoid this.

When the pipe has been cut, use a file to get rid of any 'burr' on the ends, and also to make any final adjustments needed to ensure it's square.

Joining Pipe
Supply pipes may need to be joined – eg to make two runs long enough, or to fit a branch line at an angle to a main run. Two types of joint are used – *compression joints* and *capillary joints*. The compression joint is easier for DIY use, but it is bulkier and more expensive (though this is not so important for small jobs).

Compression Joints

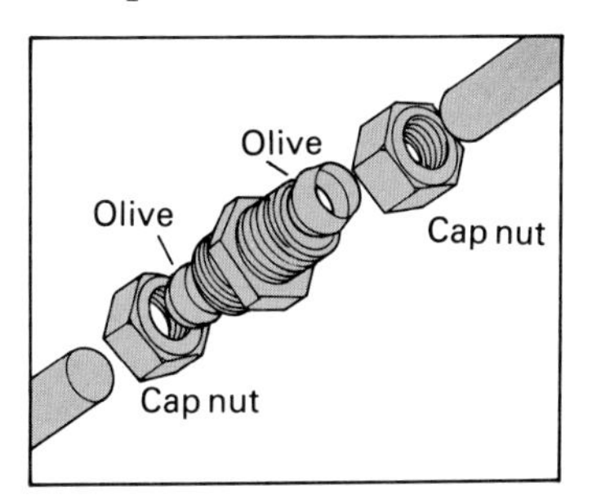

A compression joint consists of a threaded body, a small copper ring (called an *olive*) and a *nut* (cap nut). Slip first the nut and then the olive on to the end of the pipe. The chamfers of the olive are unequal; the longest should point towards the body.

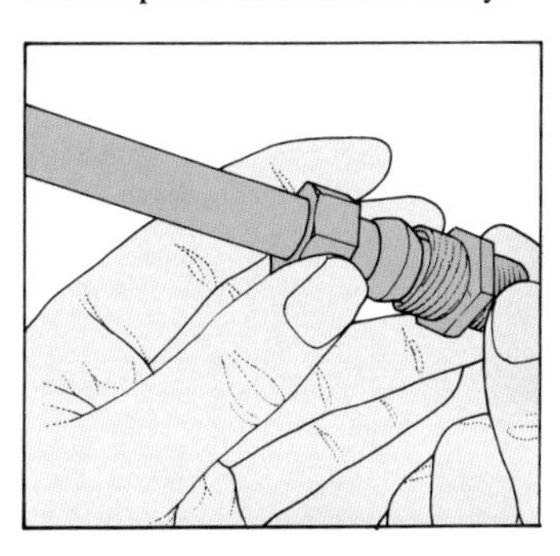

Insert the end of the pipe into the body as far as it will go, and slide the olive and then the nut to the body. Tighten the nut. The action of doing this crushes the olive, making the joint watertight.

You must tighten the nut by just the right amount. Too little and the olive will not be crushed sufficiently; too tight and it will be crushed too much. Either way, the joint will leak. To avoid it, tighten the cap nut as far as you can by hand. Now give it one complete turn with a spanner. When the work is completed and the water supply restored, inspect each joint in turn to see if it is watertight. If it weeps slightly, give the nut a series of quarter turns with a spanner, until no more water oozes out. You can, if you wish, use a jointing agent (see *Threaded Joints*) to help ensure watertightness, although the makers of joints always insist this isn't necessary.

Capillary Joints

These work by means of solder, a small ring of which is fixed inside the joint.

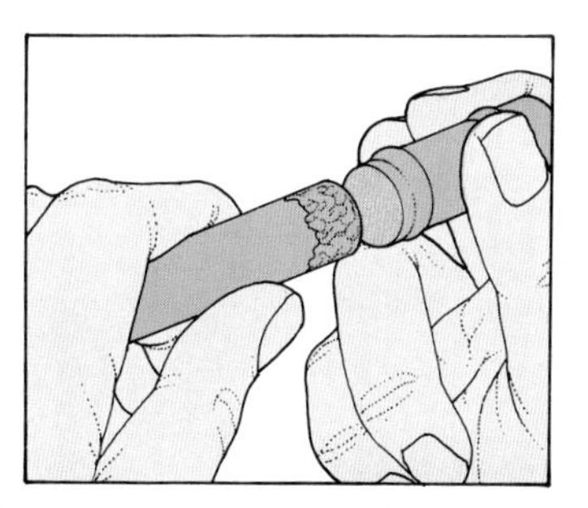

Apply flux to the end of the pipe, push the pipe fully home into the joint and twist it around so the flux is evenly spread. Now play a blowlamp flame on the joint. The heat will melt the solder, which will flow out and surround the end of the pipe, making the joint watertight.

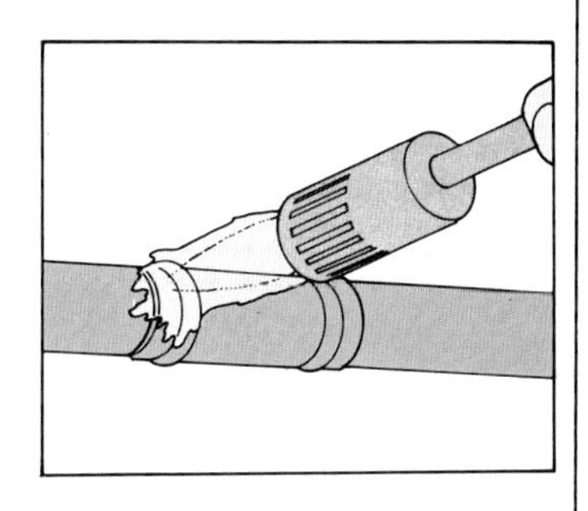

There are, however, a few points to watch. Everything must be spotlessly clean, so rub both the end of the pipe and the inside of the joint (don't disturb the ring of solder) with fine steel wool until the copper is really burnished. Then take care not to touch the cleaned parts with your hand.

Secondly, it is impossible to apply the flame to one end of a joint without the heat flowing along to the other and melting the solder there, too. So, all ends of a capillary joint must be made up at one go.

Thirdly, if you apply heat for too long, the solder will flow out too far; too little and it won't flow far enough. The trick is to

When testing a joint for weeping, don't use your fingers; skin is always moist anyway. Instead, wipe the joint with a tissue and look to see if it's damp.

Sometimes, one end of a capillary joint is defective, and doesn't have enough solder. You should have a small stick of solder handy in case this happens.

watch the joint you're heating very carefully, and remove the flame the instant a ring of bright solder appears. Then allow the solder to cool off and harden before disturbing the joint.

If a capillary joint weeps when you restore the water supply, you will have to drain off, allow it to dry (solder won't take on wet surfaces), and then reheat the joint and feed in more solder. If this doesn't work, you will have to break the joint, throw it away and fit another. This is another example of how much more convenient compression joints are.

Threaded Joints

These are used mainly for connecting pipes to taps and boilers, but you do come across them elsewhere.

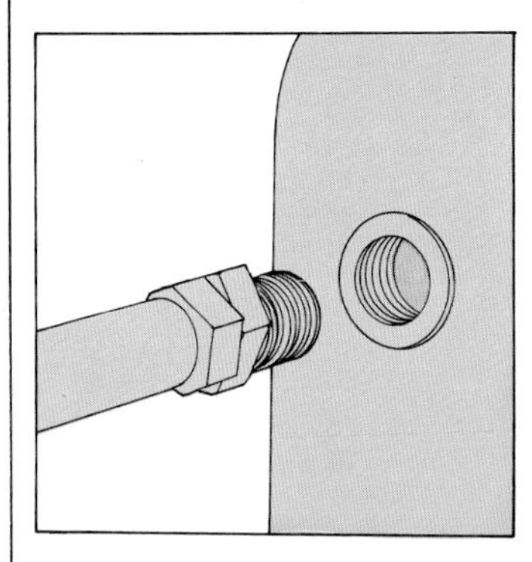

Threaded joints consist of a threaded fitting – the male part – that is inserted into a threaded hole – the female part. The joint has to be sealed with PTFE tape, or plumber's hemp, and a jointing compound, such as Boss White.

Dealing first with plumber's hemp (it is not as neat as PTFE tape, but it is more effective on the deeper threads you find on boilers).

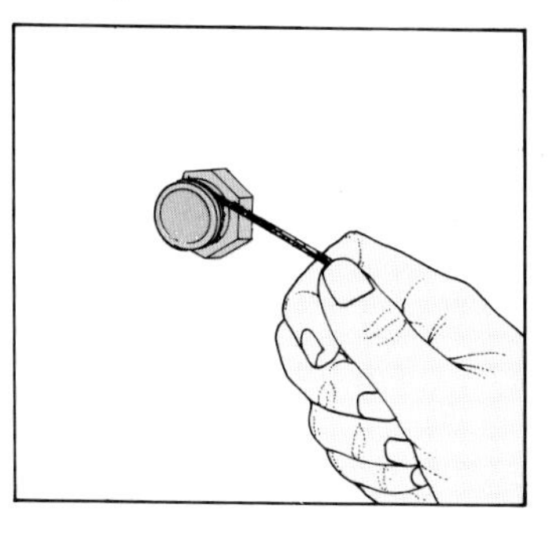

Smear a little jointing compound on the male thread and then pull off a few strands, and wind them around the threads of the male fitting in the direction that it will turn when it's being tightened up. This is to ensure it is pulled into the fitting rather than scraped off during tightening. Smear a little more jointing compound onto the hemp, insert the fitting into the female, and tighten up as hard as you can with a spanner.

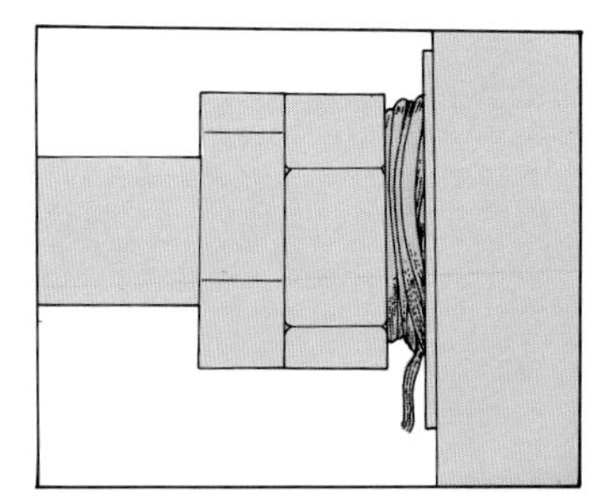

The hemp should have travelled well into the female fitting. If it's piled up outside, you have put on too much hemp, or you have wound it in the wrong direction, or perhaps not wound it tight enough. In any case, you must break the joint (unscrew the fitting) and start again.

PTFE tape looks neater than hemp, and is also easier to use. It comes in rolls like sticking plaster, although the tape itself isn't sticky.

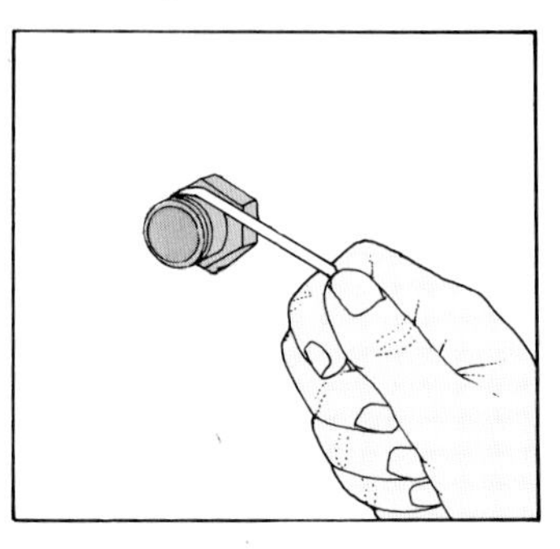

Wind it on in the same way (and same direction) as for hemp, two or three times around the fitting, allowing it to overlap. Smear jointing compound on the female thread, insert the male and tighten.

With both hemp and tape, some thread should still be visible after tightening. If not, the joint will probably not be watertight and you'll have to start again, adding more tape or hemp.

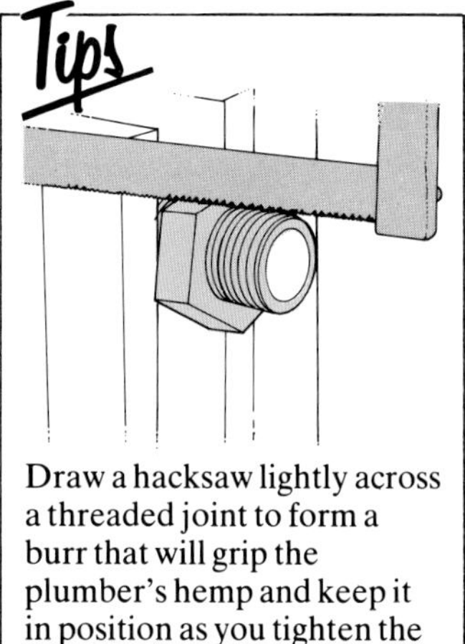

Draw a hacksaw lightly across a threaded joint to form a burr that will grip the plumber's hemp and keep it in position as you tighten the joint.

Bending

One way to take a pipe around a corner is by bending it. Plumbers have bending machines, but it's hardly worth buying one for DIY. You can bend 15 and 22mm pipe by hand.

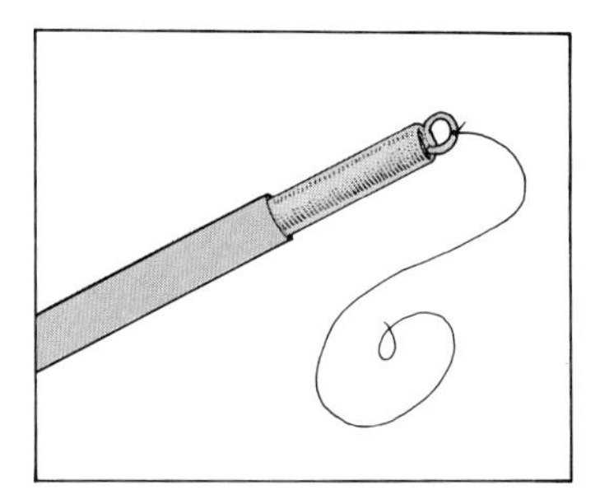

To stop the pipe from flattening, use a *bending spring*. These are available from hire shops, and you should specify one to match the diameter of your pipe.

Smear a little oil or grease on the spring, tie a length of string to one end of it and then push the other end into the pipe until the halfway point of the spring is roughly at the centre of the required bend.

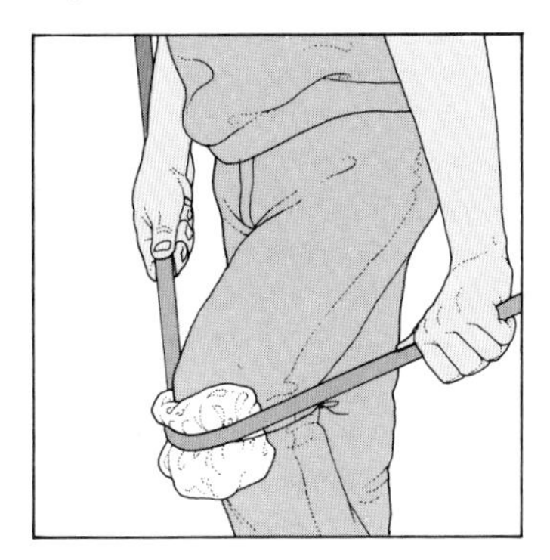

Grip each end of the pipe, place it across your knee, and give a sharp tug. If this hurts your knee, tie a protective rag pad to it.

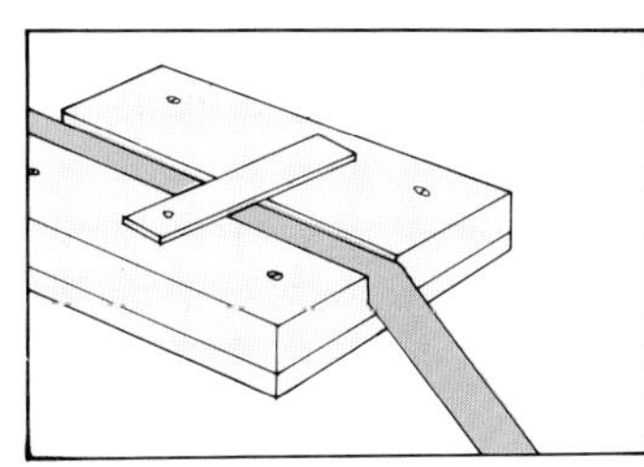

Another way to bend piping, is to use a wooden former. Nail two lengths of batten to a piece of board, the gap between them equal to the diameter of the pipe. Hold the pipe between the battens with one hand, and pull the free end to the required shape with the other.

As it is impossible to make a bend near the end of a length of pipe with either of these methods, you must bend the pipe before cutting it to length.

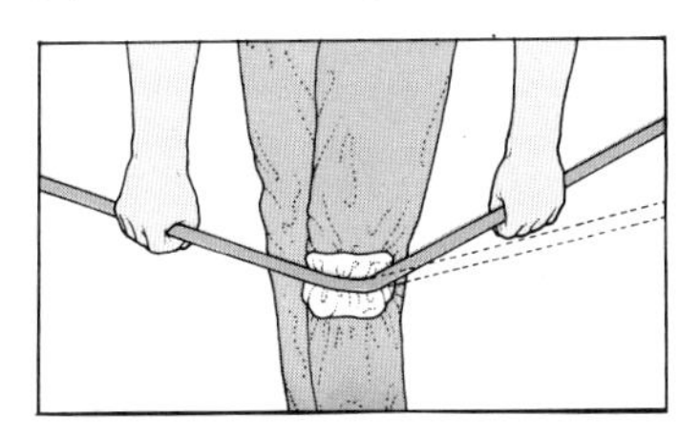

Bend the pipe slightly more than you need, then bend it back to the required angle. Twist the spring clockwise, and you should then be able to pull it out.

28mm pipe is too rigid to be bent by hand, so a bending machine must be used. However, few amateur plumbers need to use this diameter pipe – or to put bends in it. If you do, the answer is to use *elbow couplings*. These are not used generally because of their cost, and the fact that any joint introduces the possibility of a leak. However, for a small job that requires few bends, use elbows on any size of pipe.

Support

Long runs of supply pipe need to be supported.

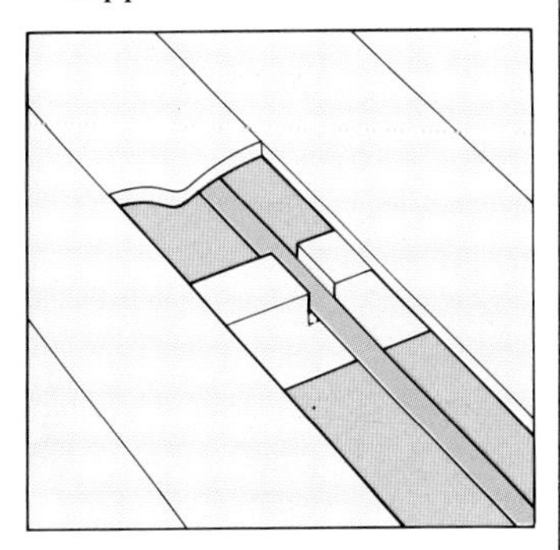

Any pipes that run beneath a timber floor, in the same direction as the floorboards, can sit in notches cut in the tops of the joists. These notches should be positioned directly under the middle of a floorboard, so that nails driven in the board won't pierce the pipe.

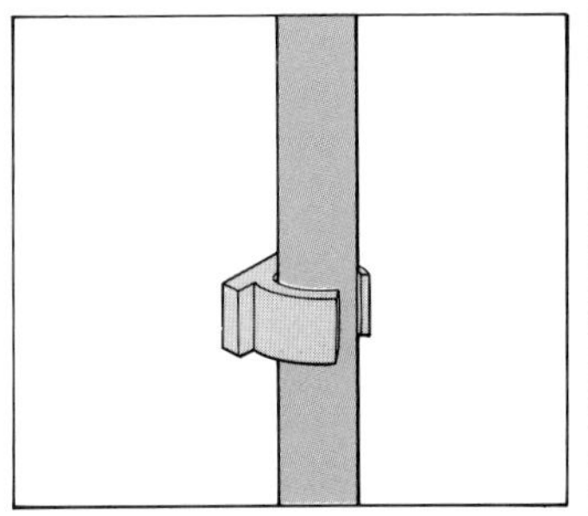

Pipes above the floor should be held to a wall or skirting board by clips and this applies to pipes under the floor at right angles to the floorboards. Use stand-off clips that space the pipe clear of the wall, rather than the saddle type that hold it close against it.

The recommended spacings between clips on 15mm pipe are 1.8m on a vertical run and 1.2m on a horizontal run; for 22 and 28mm pipe, they are 2.4m and 1.2m respectively.

INSTALLING A WASTE SYSTEM

When you are installing a new fitting or appliance, it will require a waste system. Modern waste systems are always in plastic. Four plastics are used – **UPVC** (unplasticised polyvinyl chloride), **MPVC** (modified polyvinyl chloride), **ABS** (acrylonitrile butadiene styrene), and **PP** (polypropelene).

There are two main sizes of plastic waste pipe for sinks and baths – 43mm (1½ inch) and 36mm (1¼ inch) – and as with supply pipes, the imperial dimension is the internal diameter, and the metric, the external.

Like copper, these need to be cut to length, and sometimes to be bent, although, even more than with supply pipes, it is more convenient to use an elbow joint.

Cutting Plastic Waste Pipe

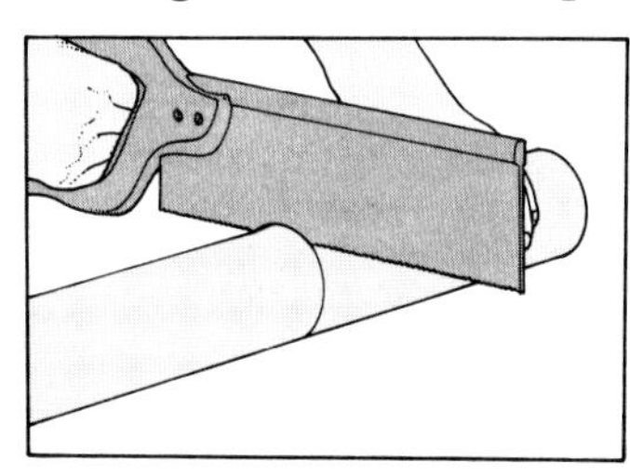

Plastic pipe is cut with a saw. This can be a hacksaw but, to help keep the end square, a fine-toothed general purpose saw is better. Measure carefully the length you need, and be sure to allow for the overlap inside any joint.

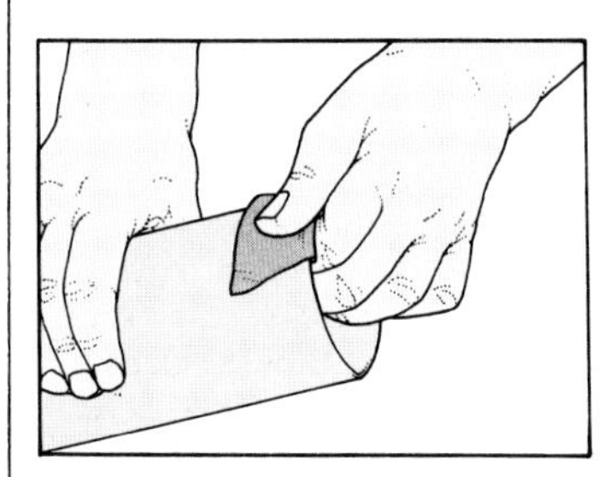

After sawing, clean off the burr, inside and out, with fine glasspaper, and then file the end of the pipe to a 45° chamfer all round.

Joints

Two methods of jointing are used: the *push-fit* and *solvent* welding. However, PP can't be solvent welded.

Manufacturers' instructions vary, so it's best to stick to one brand and follow the instructions exactly. In this section, however, we give a general guide.

Solvent welded joints

The joints used in solvent welding are smaller and neater than the push-fit kind. However, they cannot compensate for the expansion of the pipe caused when hot water passes through it. A 4m length of PVC will expand by 13mm when heated above 20°C so, on long runs, you must introduce expansion couplings at 1.8m intervals. Such long lengths rarely happen in homes. Push-fit joints allow for expansion.

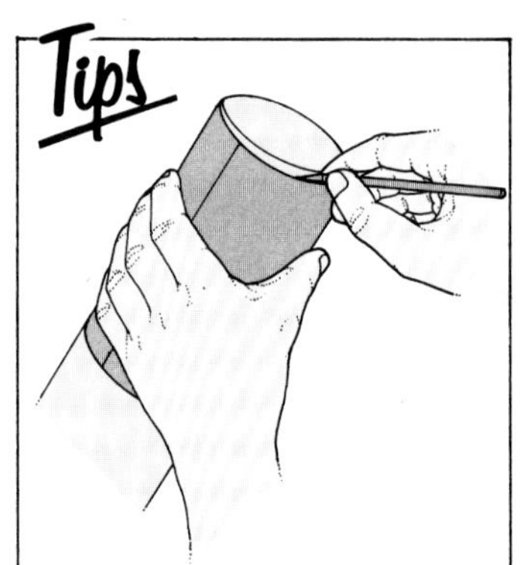

Test the cut end of a pipe for squareness by folding a length of paper around it and joining the ends. Mark any protruding edge and use a file to remove it.

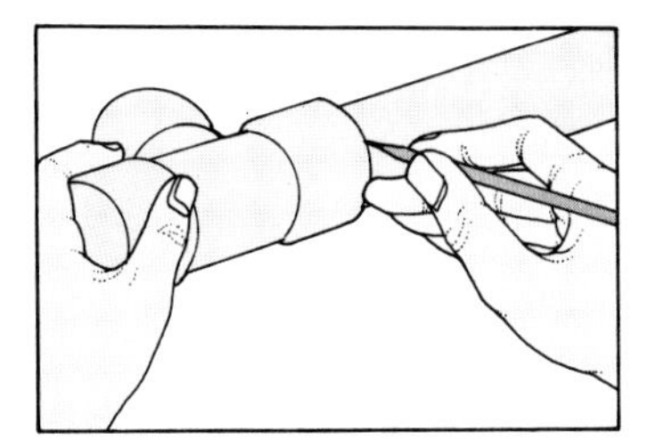

Push the end of the pipe into the socket of the joint as far as it will go, and draw a pencil line all around it. Withdraw the pipe and you will see at a glance how much fits inside the joint. Roughen this section with a file, and the inside of the socket with fine glasspaper, to provide a key for the solvent. Clean the end of the pipe and inside the socket with a spirit recommended by the manufacturer.

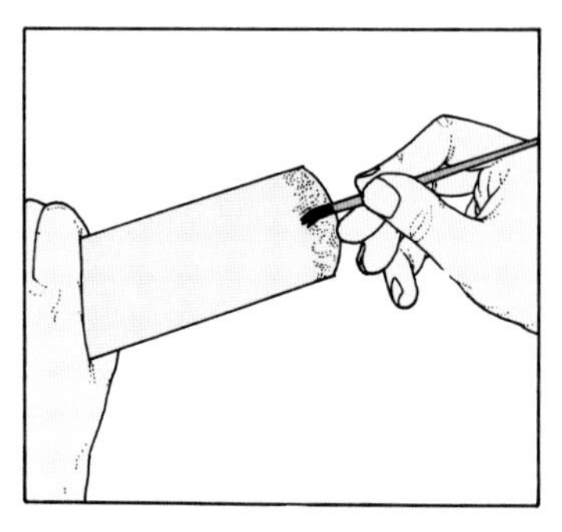

Apply the recommended solvent cement with a clean paintbrush, both to the end of the pipe and the inside of the socket. Brush it on in the direction of the length of the pipe, and make sure the whole surface is covered.

Now push the pipe into the socket (some manufacturers recommend doing this with a slight twisting action), and hold the joint securely for about 15 seconds. Wipe off any solvent that has oozed out and let the joint set for 24 hours before allowing hot water to pass through it.

Push-fit joints

Push-fit joints work by means of a rubber sealing ring inside the body of the joint. With most brands, the ring is fitted during manufacture, but it might come separately for you to fit yourself. In either case, make sure the ring is seated properly.

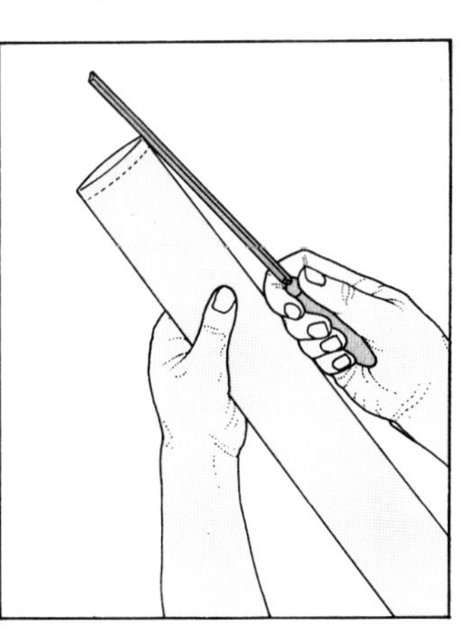

Draw a line around the pipe about 10mm from the end and chamfer back to this line with a file. Clean the end of the pipe and the inside of the socket, and smear petroleum jelly (or a silicone jelly recommended by the manufacturer) on the end of the pipe.

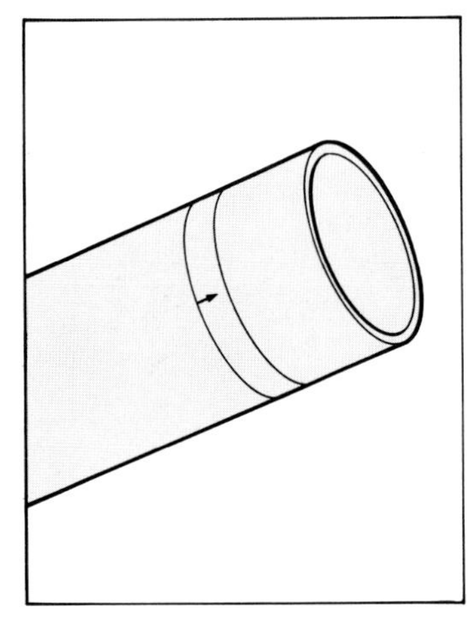

Push the pipe home and make a pencil mark where it meets the edge of the fitting. Then withdraw it 10mm as an expansion allowance.

This will ensure that the joint is watertight.

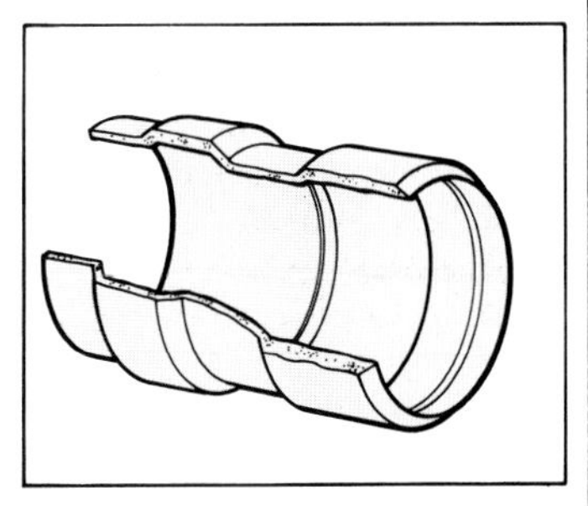

The expansion couplings that you have to fit in long runs of pipe joined with solvent cement method, have a solvent weld joint at one end and a ring seal at the other.

Support

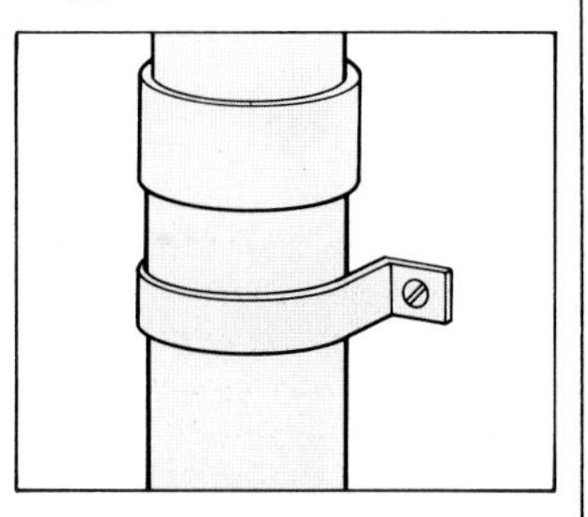

Plastic waste pipe needs to be supported by clips. These should be fitted at least every 750mm.

In modern plastic waste systems, the trap is held in place by large nuts. A washer is fitted to make a watertight seal. If a leak develops there, place a bucket underneath to catch spillages, and unscrew the nut(s). Take off the old washer and replace with a new – you might have to take the existing one to the shop.

Connecting to the Drains

You will have to make arrangements for the waste pipe to discharge water to the drains properly.

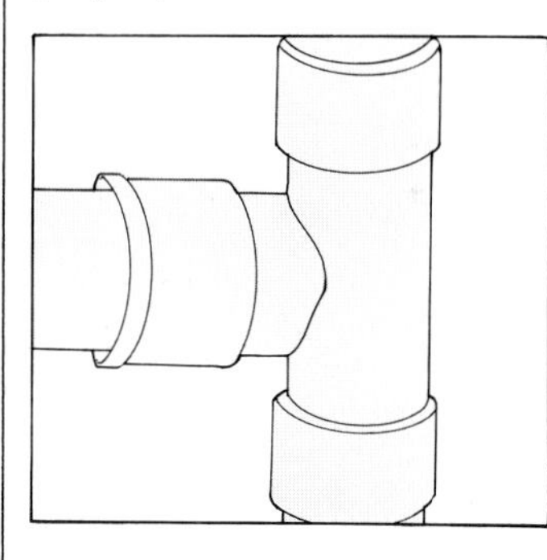

If there is an existing waste system nearby, you can often connect up the new pipe to it by means of a T fitting. Where this isn't possible, you must adopt other methods.

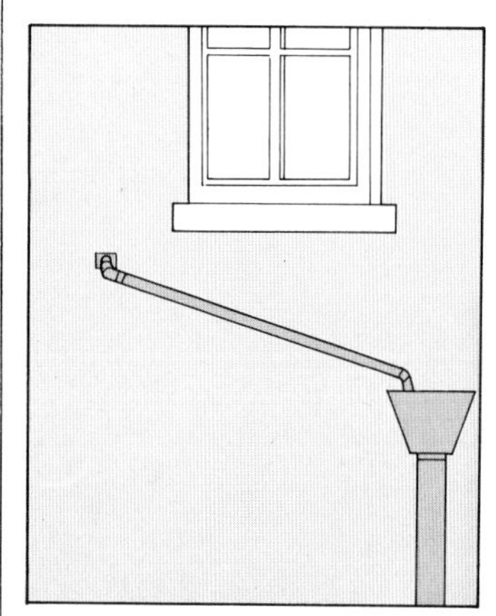

If you have the open gulley type of drainage, the waste pipe can pass through the house wall to reach it. The necessary hole should be made with a club hammer and cold chisel.

You will find it best to work from both sides of the wall. Begin gently on the plaster on the inside to avoid damage, and take care to clear out all the rubble from the hole as you break through cavity walls. Otherwise, debris might fall into the cavity and lodge, making a bridge that will lead damp from the outer to the inner leaf.

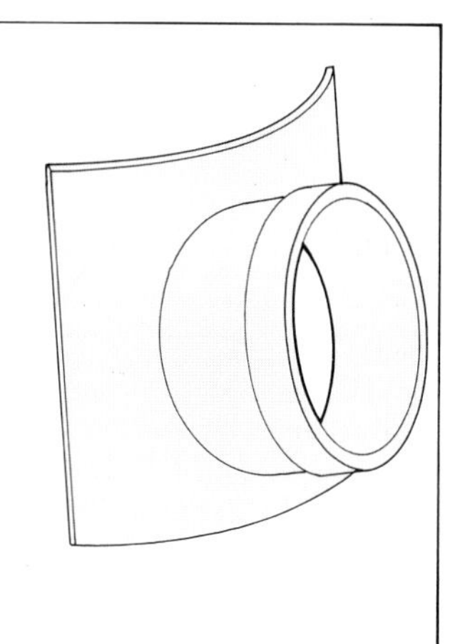

If your house has the one-pipe drainage (single stack) system, your new waste can be connected to the stack. The connection is made by means of a special device, known as a *boss*.

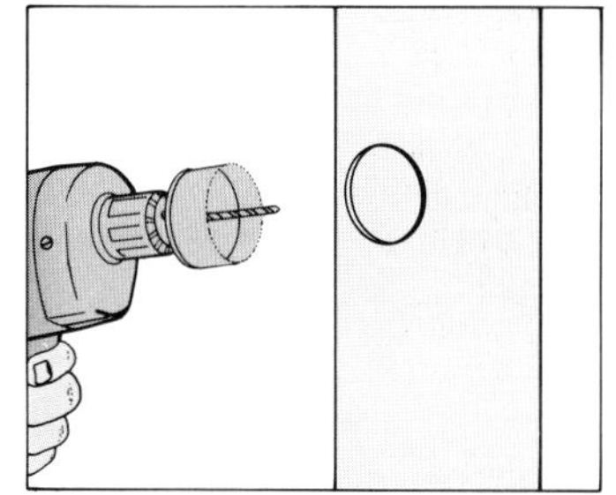

A hole must be cut in the stack using a hole saw (you can usually hire it) fitted to a hand or electric drill. The hole must not be within 200mm of a wc connection on the opposite side.

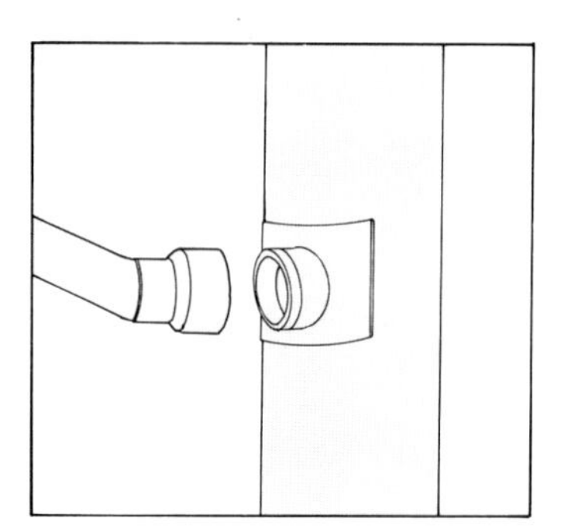

Boss

Various types of boss are made, most needing solvent welding to the stack. Then the waste pipe is connected to the boss by either a ring seal or solvent weld joint. Ask when you buy it, to make sure you get the one suitable for your needs.

One big problem with single stack systems is that water rushing down it can cause siphonage (suction) that can empty the water out of the waste traps.

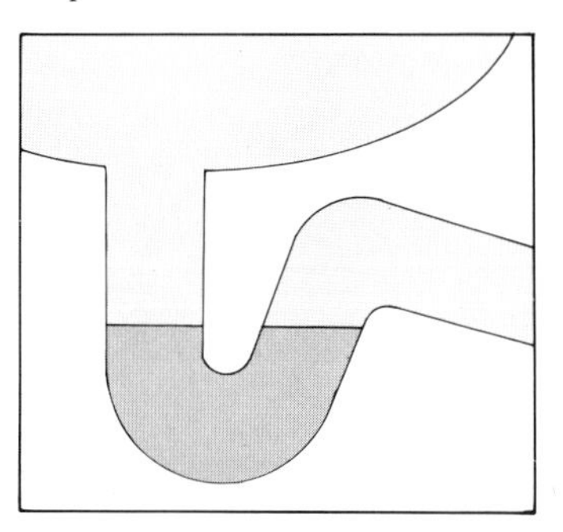

To prevent this, deep seal 75mm traps must be used, and the waste pipes laid with as short a run as possible and very shallow falls. In the case of washbasins, it may be necessary to connect a 38mm waste (instead of the usual 32mm) or fit anti-siphon traps. All these are points to be discussed with your local authority.

Installing taps

There are more taps in a plumbing system than the familiar sink, basin and bath taps. *Stop taps* positioned at strategic points will allow you to isolate various appliances when maintenance is called for, without shutting down the whole system. There should be stop taps in the supply pipes from the cold storage tank (to save draining it during maintenance) and, if there is a separate pipe to supply the hot cylinder, it's very handy to be able to turn it off separately, too. Few homes have enough; it is a good idea to fit extra ones.

Stop Taps

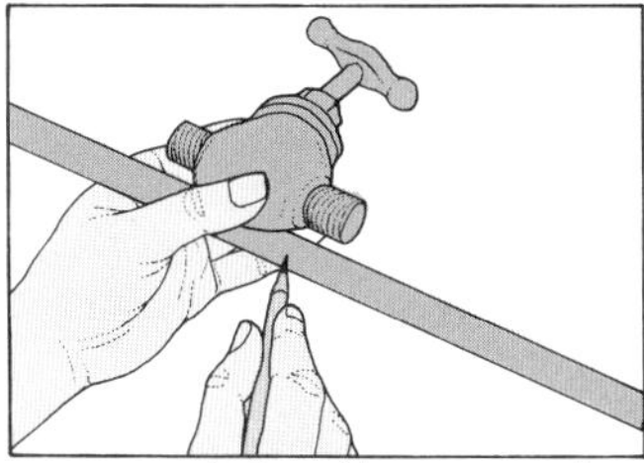

A stop tap is fitted by means of a compression joint at each end. Select a suitable position for it, and measure how much of the supply pipe will have to be removed to accommodate it. The pipe will have to fit right up to the shoulder inside the joint socket, so measure inside there.

Turn off the supply. Cut out enough pipe to take the stop tap. There probably won't be enough room to wield a pipe cutter, so use a hacksaw, taking care not to flatten the ends of the pipe. Clean up the ends and slide on the nuts and olives.

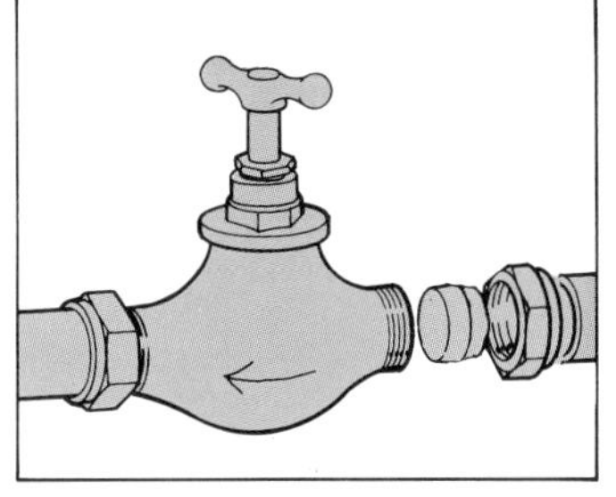

Place the tap in position with the arrow on its body pointing in the same direction as the water will flow. You will be able to spring the pipes far enough apart to do this.

Tighten the compression joint at each end of the tap, restore the supply and check that the fitting is watertight.

Mixer Taps

Mixer taps can be fitted in place of two single taps at a kitchen sink. Or a shower fitting can take the place of bath taps. The only pre-condition is that the holes in which the present taps sit must be at a distance that allows the two 'tails' of the mixer to be inserted. In modern baths and sinks there should be no problem as the holes are at a standard distance. In any case there is often some leeway over this as the two legs of some mixers can be swivelled to vary the distance between them. Measure the distance between the centres of your single taps before you buy.

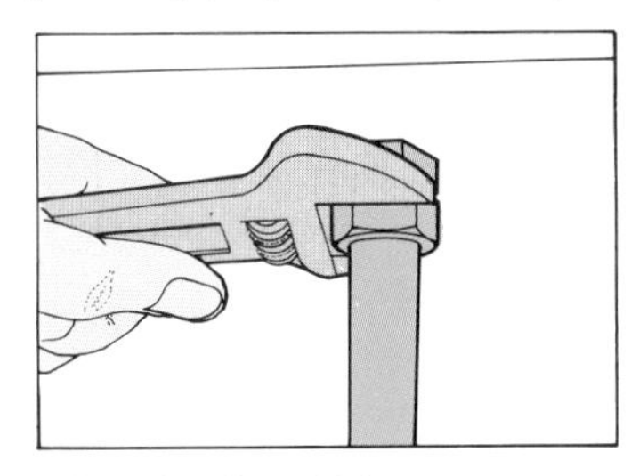

After shutting off the supply, disconnect the supply pipes from the taps – by unscrewing the cap if it's a compression fitting, or sawing through it if it's a capillary joint.

The biggest difficulty will be removing the old taps, as they may be so corroded it will be hard to loosen the nuts that hold them in place.

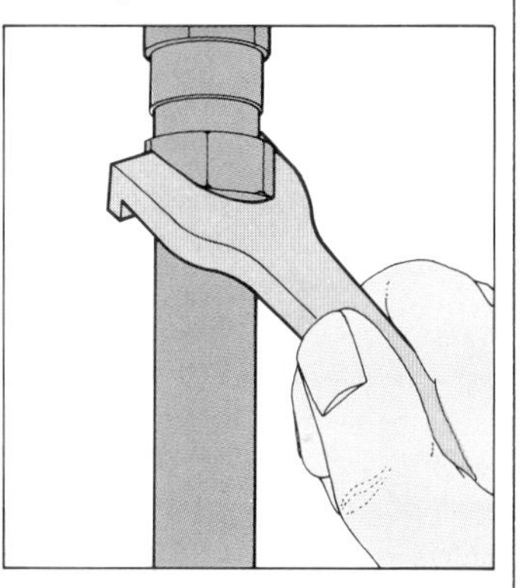

There will not be enough room to wield an ordinary spanner, so you should use the type known as a *crow's foot spanner*. If the nut still sticks, try squirting penetrating oil (or other products designed to free corroded joints) around the nut.

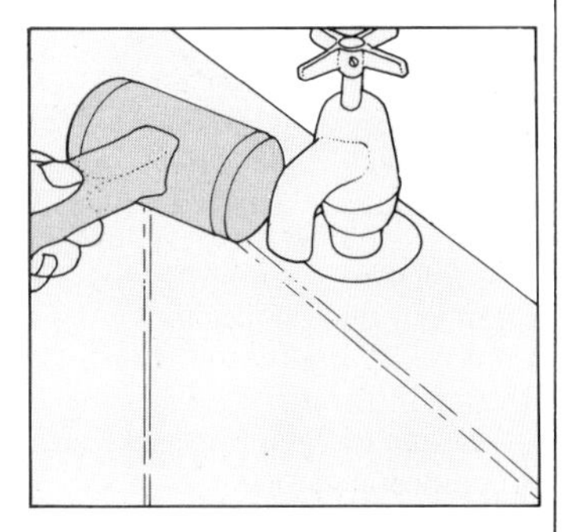

One trick that works is to get a helper to hold the tap while you turn the spanner, or hold the spanner tightly while your helper gently taps the spout of the tap to unscrew it. Be careful not to

Sinks & basins

damage the sink or bath.

With the old taps out of the way, fitting the new is easy.

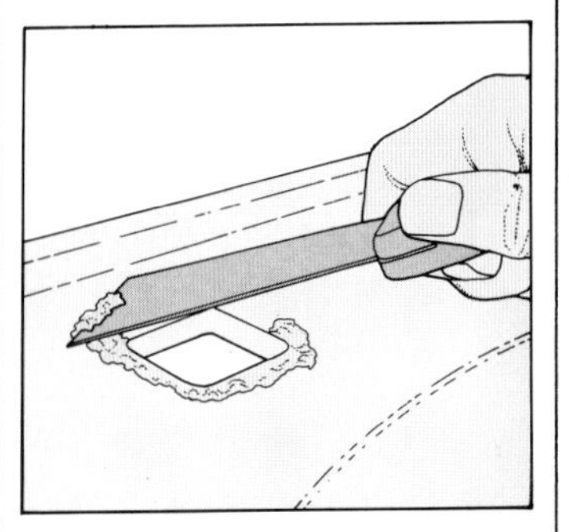

There should be full instructions, but usually the tap legs are bedded in putty (scrape off the old putty first). There is a washer underneath and a back nut holding everything in place. Sometimes a washer is fitted both above and below, and there may be a second locking nut.

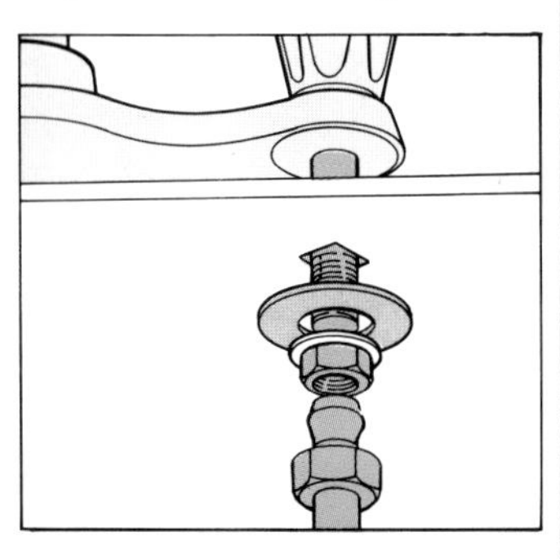

The mixer is connected to the supply by means of a *tap connector*. This fits to the tail of the tap by means of a threaded joint, and to the supply pipe with a compression joint.

New Taps

New taps for old are an easy way of modernising an old washbasin or bath. The procedure for fitting them is the same as for a mixer.

A basin can be sited where none is fitted at present (for instance a bedroom) provided it can be conveniently connected to the drains.

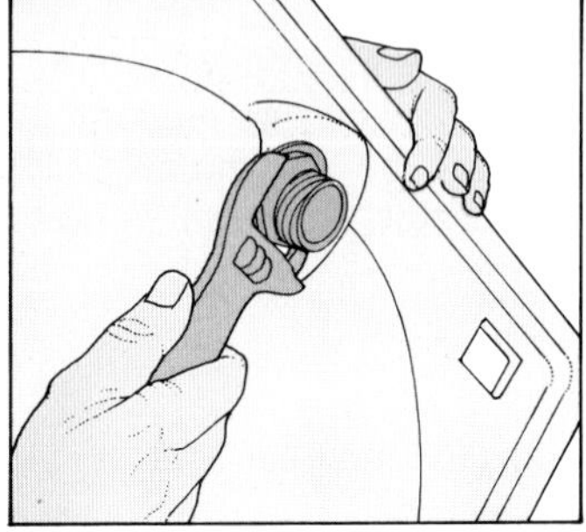

To install a new washbasin prepare it by installing the waste outlet fitting, which is held in place by a back nut underneath the basin, and made waterproof with the washers supplied, or mastic. Do not over-tighten the nut, or you may damage the basin.

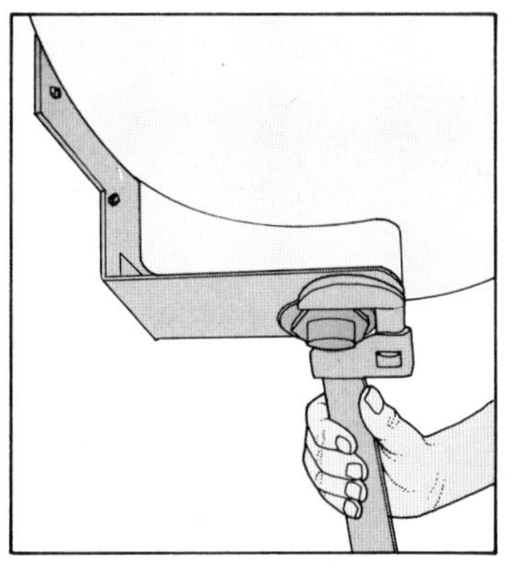

Fix it to the wall with brackets of the correct size, or on its pedestal according to the instructions supplied with it.

If the basin is to be set into a worktop, use the template (supplied by the manufacturer or make one yourself from card) to mark, then saw out, a hole to receive it. Mastic is used to make a waterproof seal between basin and worktop.

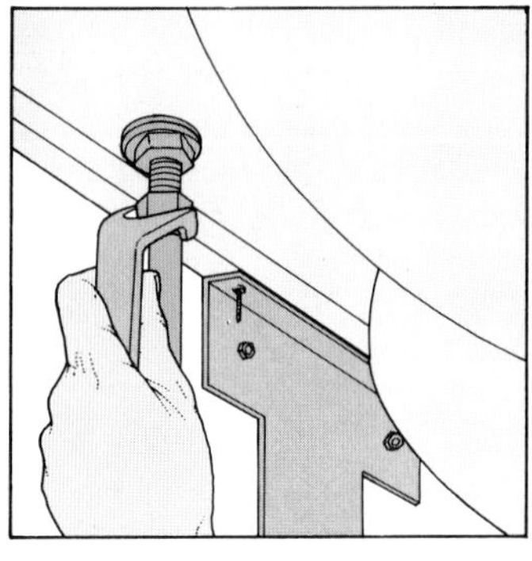

Fit taps to the basin, tee into the hot and cold supply at a convenient point, and connect supply pipes to the taps.

Connect a trap to the waste outlet, and run a plastic waste pipe to discharge into a convenient drain.

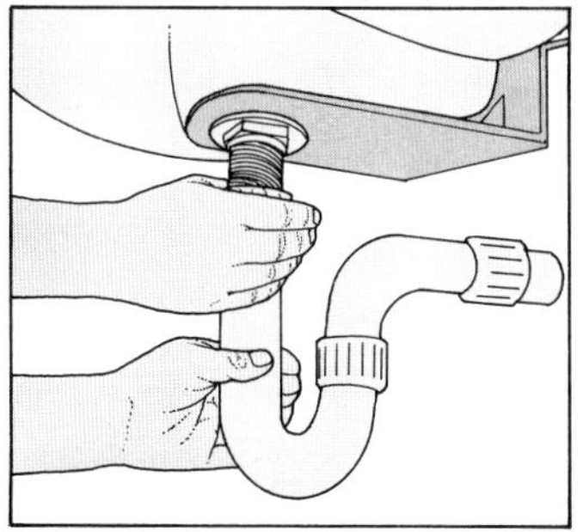

If the basin is to take the place of an existing one, turn off the supply and disconnect or saw through the supply pipes to its taps. Disconnect the trap, if the waste is plastic, or saw through it if it's metal.

Lift the basin clear, and remove the pedestal or brackets. Replace a metal waste system with a new plastic one.

Sinks are dealt with in a similar manner.

Washing machines

Some machines need both a hot and a cold supply; others only cold. Follow the maker's instructions. There are two ways of connecting up to the supply.

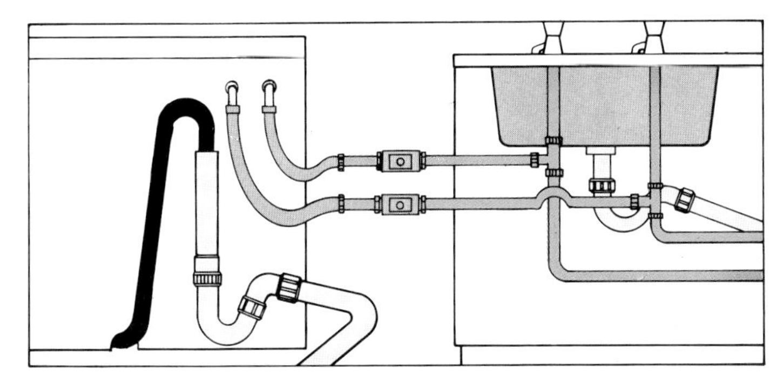

Supply

Find a convenient point, cut into the supply pipe, and fit a T junction. To the branch of the T, connect a short length of pipe and either an ordinary stop tap or one purpose-made for washing machines. The stop tap is essential in case the machine has to be removed for maintenance, or a leak develops in the hose or machine itself. Some manufacturers recommend that the supply is turned off at the stop tap whenever the machine is not working. At the end of the pipe, fit the coupling to which the flexible hose of the machine can be connected. This is the method professional plumbers usually adopt.

Do-it-yourselfers may prefer a simpler method: you can use what is called a *self-cutting tap*, made especially for plumbing-in appliances.

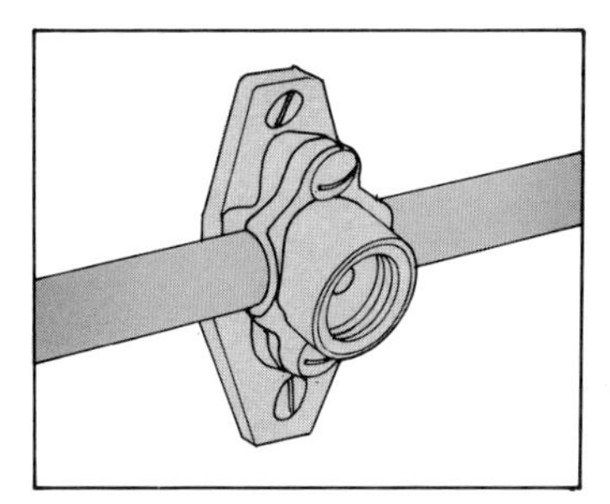

First fit the tap's *saddle clamp* at a convenient point on the supply pipe, and tighten the screw that holds it in place.

On the side of the clamp is a threaded socket, into which you screw the main body of the tap. This has a special tip that cuts into the pipe, so connecting with it. No water can escape during the fitting, so you don't even need to turn off the supply. Tighten the tap with a spanner, and you have your stop tap ready to connect to the supply hose of the machine.

Waste

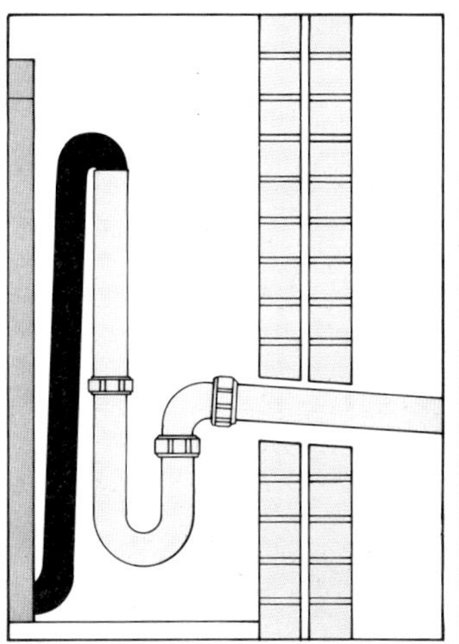

Construct the machine waste system in pipe of the diameter recommended by the manufacturer. The machine has a flexible hose waste pipe that has a bend at the end of it. This end hooks into a vertical pipe that is fitted near the wall close top the back of the machine. The top of the pipe must be left open, and the height above the floor will be specified in the instructions.

From the vertical pipe, a plastic waste pipe can be run to an outside gulley, or joined to the main stack (see page 16) or connected by means of a T junction to an existing waste pipe. Also, a trap must be fitted somewhere in the run.

Showers

Electric Showers

Electric showers are connected directly to the rising main and, since the pressure of the main ensures that water comes out of the rose with sufficient force, there is no need to worry about the head of water. The appliance has an instantaneous heater that warms up the water as it flows out when the tap is turned. No hot water is stored.

To install one of these showers, fix it to the wall at a suitable spot according to the manufacturer's instructions. The shower can discharge into a bath, or you can site it near a basin, so that it can be used for hair washing.

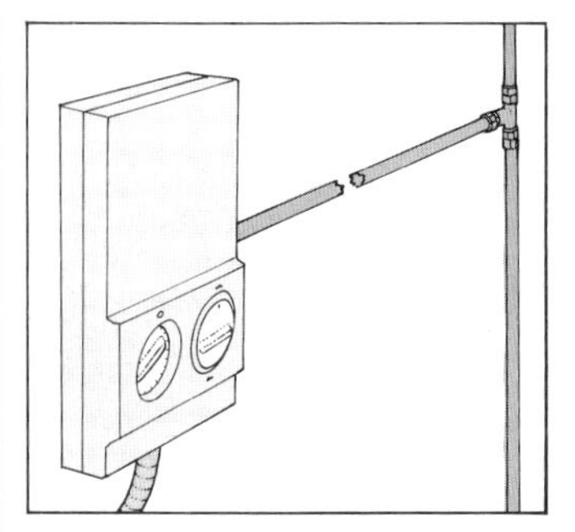

Run a supply pipe from the shower to a convenient spot on the rising main, and connect it by means of a compression T joint.

There are special regulations regarding the wiring of electric shower units. Instructions for electric wiring are given in the companion 'Do It' book, *Wire It*.

Conventional Showers

You can also install a shower which is fed from the hot cylinder and cold storage tank. This can be either in place of, or in addition to, the existing cold and hot taps on your bath.

Two basic types of shower unit are available. Firstly, there is the *thermostatically controlled unit*, which will maintain the temperature you've selected throughout your shower. These units are, however, very expensive, and most people settle for a *mechanical unit*, with which the temperature can vary.

The main disadvantage of a mechanical unit is that it may suffer from what plumbers call an *auxiliary draw-off*. This means that in the middle of your shower, someone may turn on a tap or washing machine, or even flush a wc, and so cut down the flow of water to your shower. If the tap is a hot one, it will be the hot supply that is reduced, and you will get a sudden cold douche, an unpleasant shock perhaps, but nothing else. Much more dangerous, however, is when it is the cold supply that is restricted, giving you a scalding hot shower.

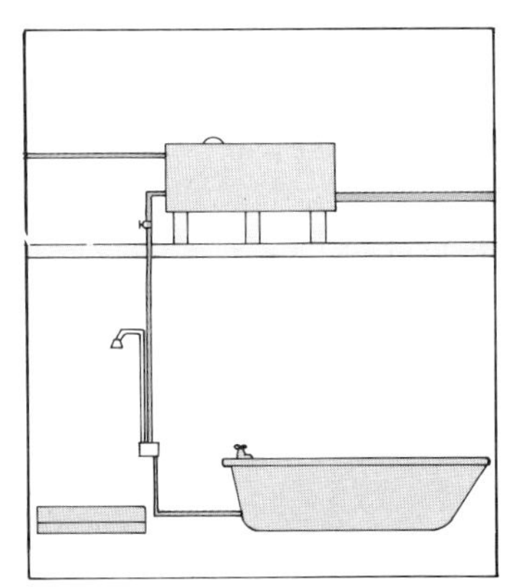

To prevent this happening, you should give the shower its own cold supply pipe directly from the cold storage tank – not joining on to a conveniently placed existing cold supply. You can arrange the hot supply by teeing into the supply to the bath's hot tap.

The other point to watch with a shower is that you must have a sufficient head of water to give enough pressure (see page 6). The maker's instructions will tell you the minimum head required (usually about 1m), but the more you can exceed the minimum, the better your shower will be. If you can't get a sufficient head of water, it is possible to boost the pressure with a pump. Most people, however, would settle for an electric instantaneous shower unit that gets its pressure from the mains.

Shower Cubicles

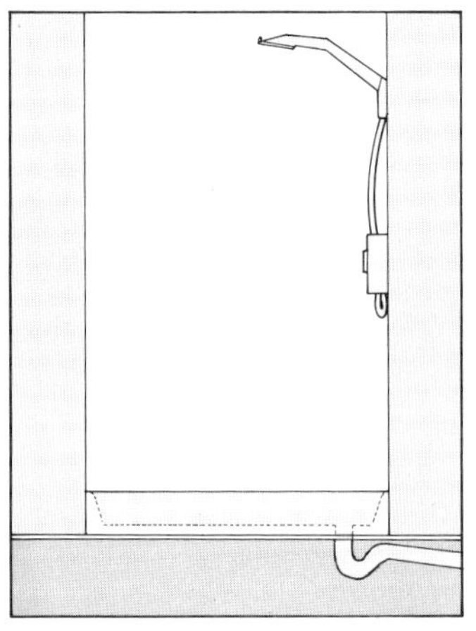

Installing a shower cubicle in a bedroom, landing or under the stairs is an inexpensive way of increasing your home's bathroom facilities. Basically, it is fitted just like a sink or basin. The big difference is that, since the waste outlet is so low down, you may have problems getting a sufficient fall in the waste run. It should be 25mm per 300mm for a short run of 600-900mm, but only 12mm per 300mm where the run will be 3-4.5m. Also, the trap may have to be fitted under the floorboards.

Right *A vanity basin fitted into a counter in any bedrom is a practical luxury.*

Far right *Designed for someone confined to a wheelchair, this simple, easy-to-maintain bathroom has all the plumbing concealed behind a tiled box. The washbasin has both taps easily accessible, and has been plumbed-in at a low level to enable it to be used from a wheelchair.*

Centre *Why not put your own bathroom at one end of your bedroom, divided off by sliding screens?*

Below left *How about fitting an extra shower under the stairs? Ideal for a sporting or gardening family, and when the door in front is closed, no one would know it's there.*

Below right *To help ease the strain on an overworked family bathroom, see if you can add a shower on your landing.*

REPAIRS AND MAINTENANCE

Quite a few things can go wrong with your home's plumbing system, and various parts can deteriorate and need renewal. Here we look at some of the more common jobs.

Taps

Taps are made of a non-rusting metal (usually brass, or chrome plated on the outside) although plastic taps are now being introduced.

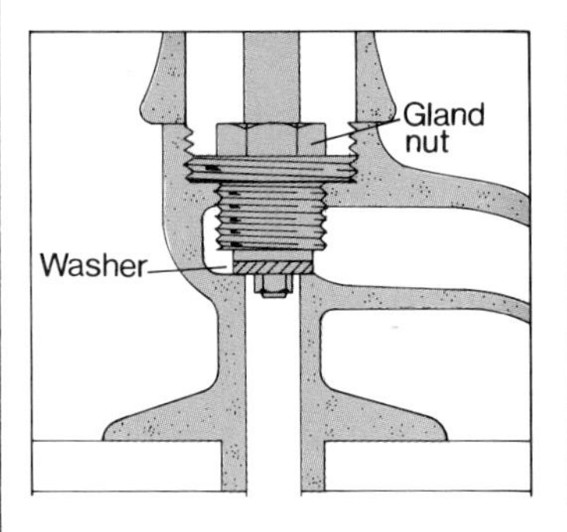

Two metal surfaces pressed together can't provide a perfect water seal, so a *washer* is used to make everything watertight. Washers were once made of leather or fibre, but nowadays are rubber or nylon.

When you close a tap really hard and it still drips slightly, the washer needs to be replaced. This is how to replace the washer on a domestic tap – and the method is the same for stop taps.

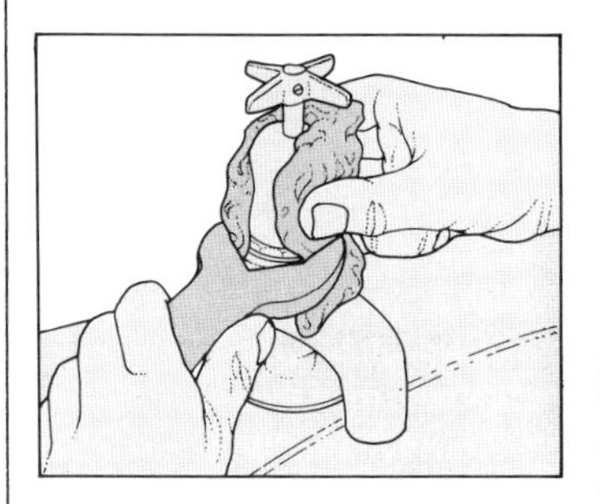

Turn off the water supply and open the tap. Then unscrew the *shield*. If corrosion stops you doing this, use a spanner, protecting the shield with a cloth. Still no success? Try pouring boiling water on the shield to make it expand.

Loosen the hexagonal head with a spanner and lift the top clear of the body. The washer is fixed to the jumper plate with a small nut. Remove the nut and replace the washer with a new one of the same diameter – the maker's name should face downwards. Reassemble the tap.

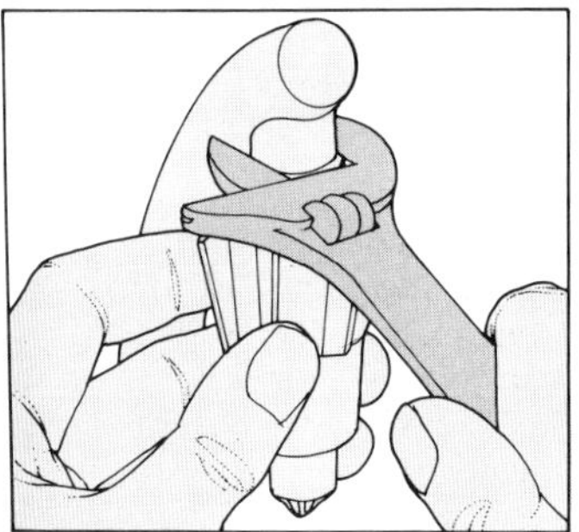

Supataps are treated differently. There is no need to turn off the supply. Loosen the gland nut with a spanner and unthread the nozzle, checking that the valve that stops the flow of water has dropped into position.

The supatap has a combined jumper and washer, housed in its *flow straightener*.

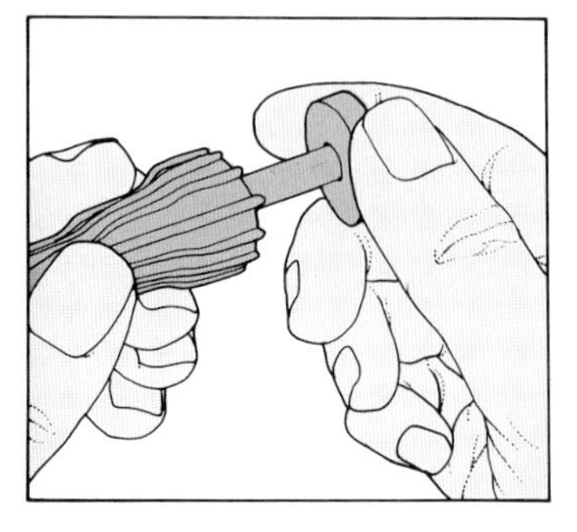

Take out the flow straightener – one way is to push it with a pencil from the other end – and replace the washer. Clean the flow straightener before reassembling the tap.

Leaks

If water oozes out from the top of the shield, the fault lies with the gland unit.

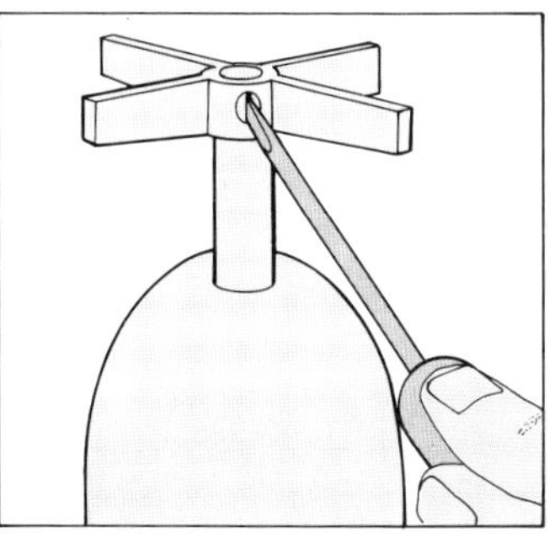

Begin by removing the top. A cross top will be fixed by a small grub screw. Remove this and place jaws of a spanner underneath and lever the top upwards.

With modern-style taps, prise out the H or C button in the middle of the top and withdraw the screw underneath. You will now be able to pull off the head.

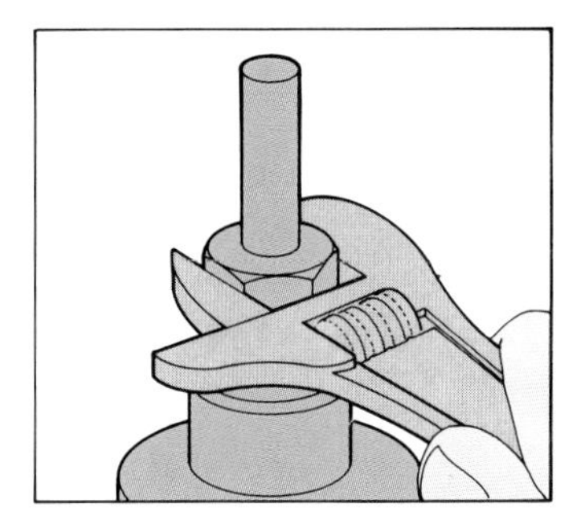

In both cases the gland nut will now be accessible. Tighten this half a turn with a spanner. Temporarily replace the head and check that the tap is easy to turn. If it's too tight, slacken the gland nut slightly.

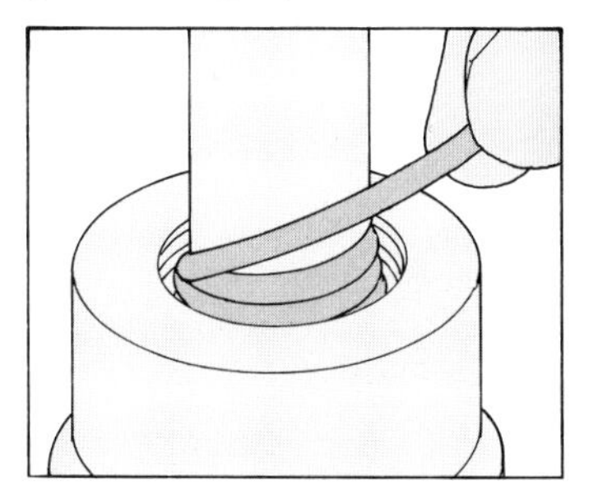

Check, too, whether the tap still leaks. If it does, remove the gland nut and re-pack with string smeared with petroleum jelly around the spindle.

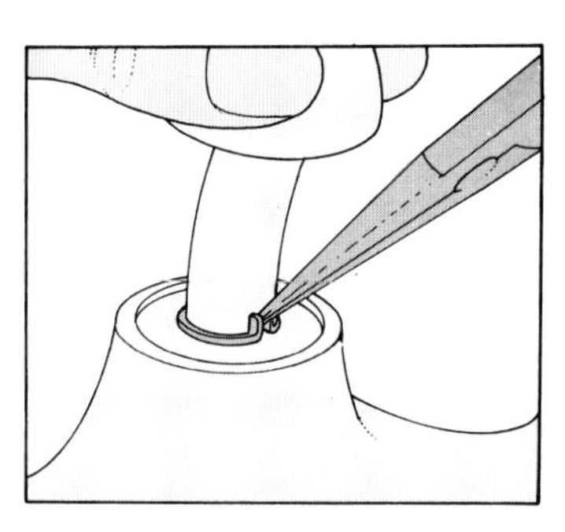

If there is a leak at the bottom of the swivel nozzle of a mixer tap, raise the shroud at the base – some you merely lever up, others have to be unscrewed.

Prise up the circlip that will now be revealed, and expand it. Slide it up the nozzle, and lift the nozzle clear.

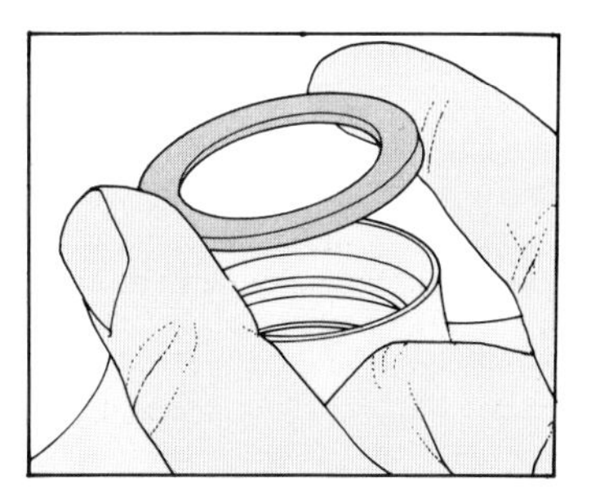

Replace the seals in the base, and wet the base before re-assembly.

Shower Mixers

If there is a drip from a shower mixer, a defective O ring is the likely cause.

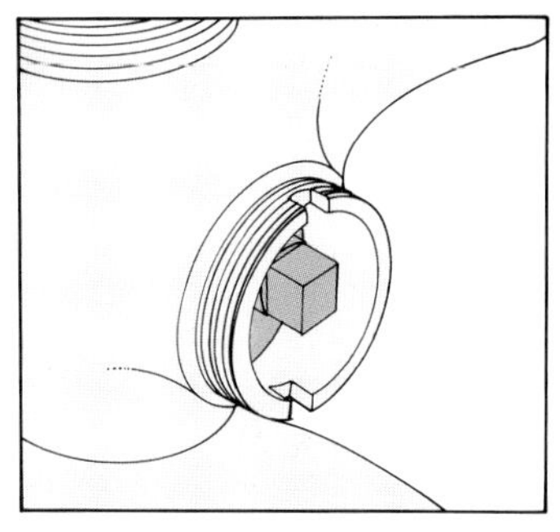

Take off the shower-bath diverter by withdrawing the screw that holds it. You will now see a slotted connector.

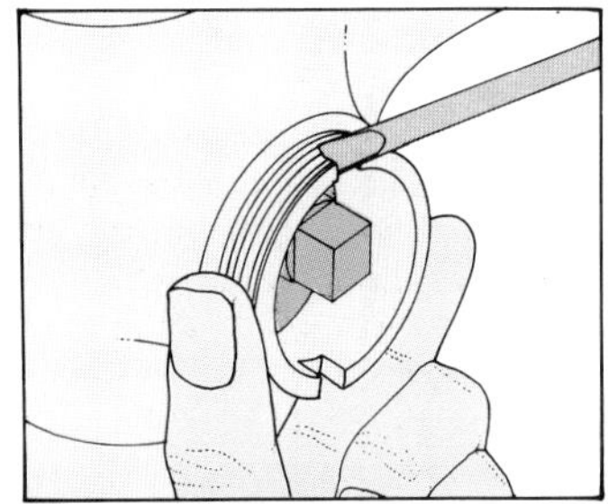

Remove this by inserting a screwdriver in the slot and pushing.

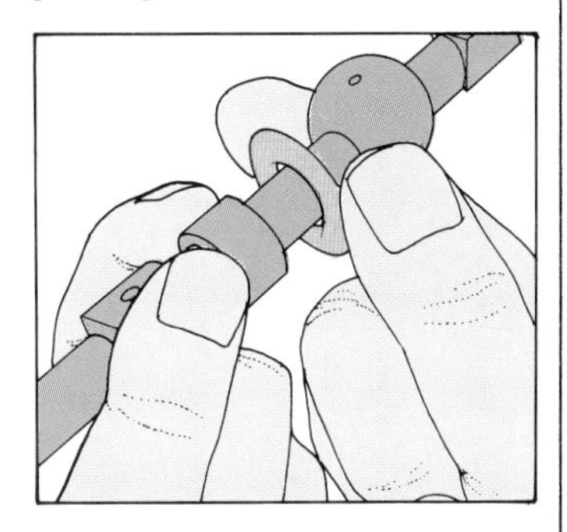

Remove the old O ring, slide on the new, and re-assemble.

Turning off the Supply

When you want to work on a tap or cistern, you must shut off the supply of water to it. When it isn't controlled by its own stop tap, this involves turning off the stop tap on the rising main (so that the whole system is closed down) and draining the cold storage tank. Before doing this, either switch off (or lower right down) the thermostat of any boiler or heater that supplies hot water. Solid fuel fires and boilers should be damped right down.

Once you have turned off the raising main, open the tap on which you want to work – or in the case of a lavatory, repeatedly flush – until no more water can be drawn. Then you can start work. Restore the water supply as soon as you've finished.

Overflowing Cisterns

A cistern overflows when the ball valve doesn't shut off the supply when the water reaches the correct level – usually about 12mm below the overflow pipe. This may be because the arm isn't set properly.

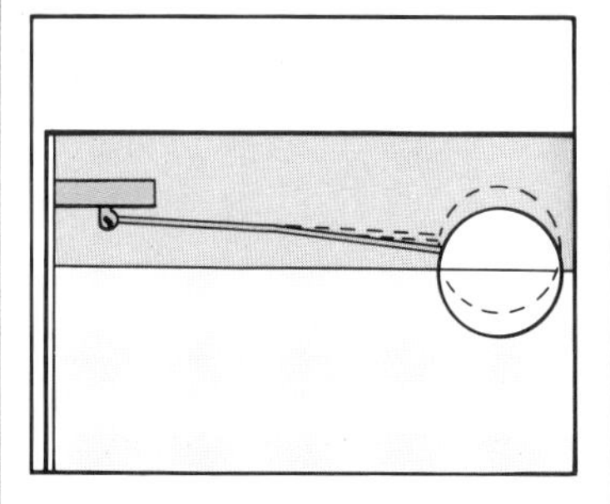

If the arm is metal, try bending it so the ball (or float) rides lower, and the valve will shut off the water before the level reaches the overflow.

If the ball is submerged, instead of floating on the surface, it is punctured or damaged in some way. Don't bother trying to mend it – you could never get rid of all the water inside. Better to replace it – a new ball is inexpensive. Simply screw the new ball onto the thread end of the arm.

You will have to shut off the water while changing the float.

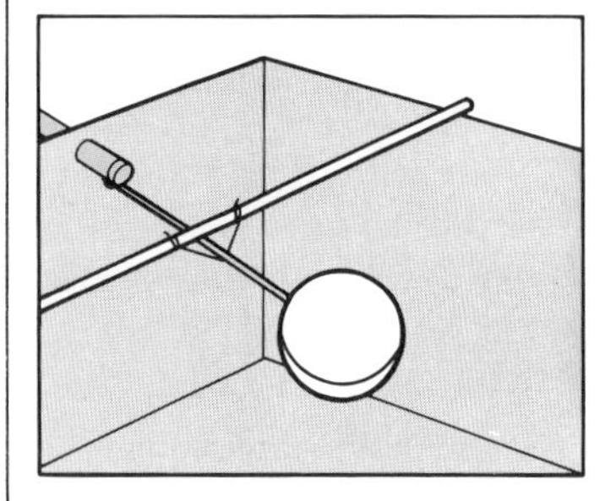

If there's no stop tap handy, place a stick across the top of the tank and loop a string around arm and stick to keep the arm raised.

There is another possibility. If the valve still drips when you raise the arm as high as it goes, then it needs a new washer.

First, turn off the supply to the valve. Several types of ball valve are fitted, but the most common is the *Portsmouth*.

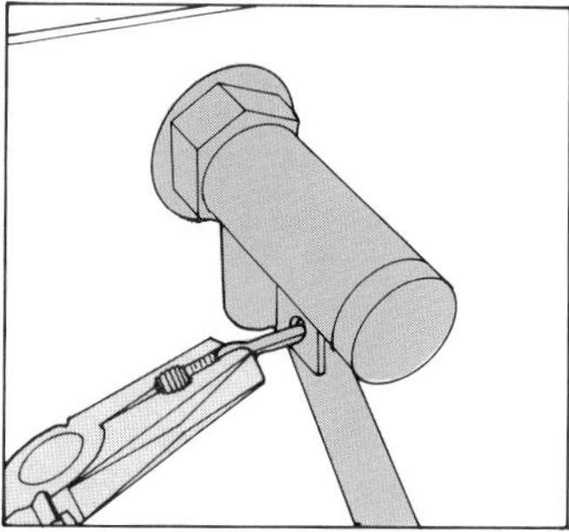

With the **standard Portsmouth,** take out the split pin holding the float arm, and remove the arm. Place a screwdriver in the slot under the valve, and lever out the piston.

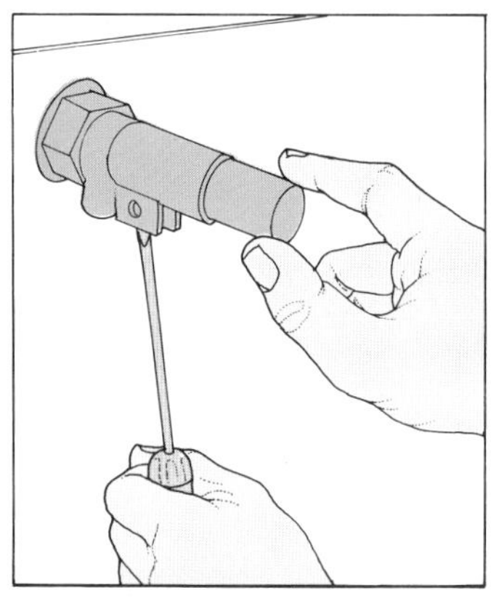

Push your screwdriver in the slot in the piston and unscrew the washer retaining cap. Fit a new washer, lightly grease the components, and reassemble.

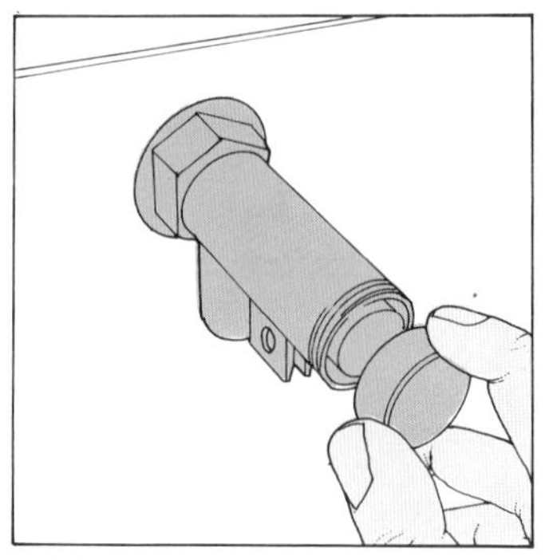

With a **Portsmouth equilibrium valve**, you have to unscrew the valve's end cap before you can remove the piston.

If fitting a new washer is not successful, the washer seating may be damaged. You can hire a re-seating tool to smooth it down, but as a last resort you may have to fit a new valve. Check with your local plumber's merchant to ensure you get one suitable for the water supply.

If you're working on an existing installation (rather than a new one) turn off the supply to the cistern and push the ball down until no more water comes out. If this could cause an overflow, lower the water level in the cistern by turning on a cold tap or, if it's a lavatory cistern, flushing the wc.

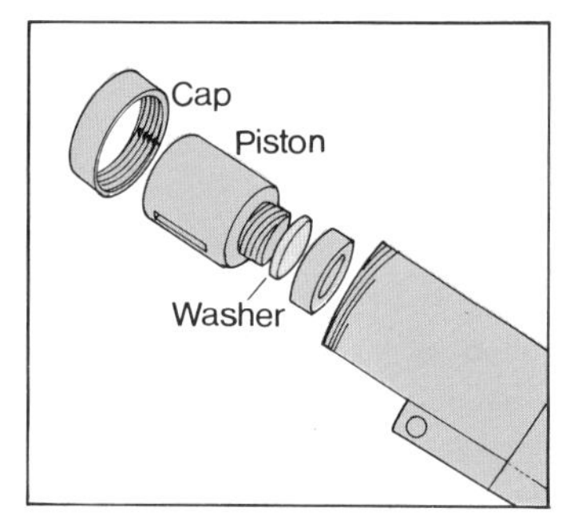

Remove the fixing nut, and take out the old ball valve. Slip one of the washers provided over the threaded tail pipe of the valve, and push the tail through the hole in the side of the tank.

Fit the other washer on the outside and tighten up the back nut. The supply pipe is then connected to the tail pipe by a compression joint.

Freeze-ups

If your supply pipes are not protected by lagging, the water in them will freeze in really cold weather. Not only does that mean the inconvenience of not being able to draw water; water expands as it turns into ice, and that can fracture a pipe or force a compression joint apart. So, when the thaw comes, you will get a flood.

As soon as you realize you have a freeze-up, try to thaw it under controlled conditions, so as to avoid a burst. Turn on the affected tap (keep a watch so the sink doesn't overflow) and work backwards from it looking for signs of the freeze-up – the pipes will feel at their coldest where the ice has formed.

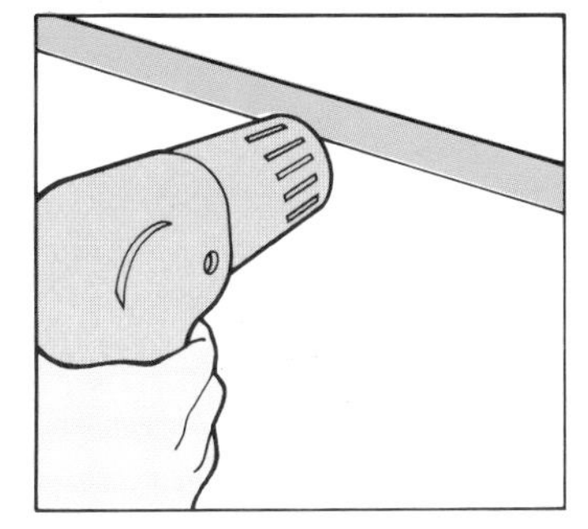

Turn off the supply and apply heat to the affected part. The most convenient way of doing so is with a hairdryer, but an electric fan heater will do. As a last resort, wrap rags soaked in warm water around the pipe. Eventually you will melt the ice, and the water will flow again. Resolve immediately to lag the pipes so that you never suffer such trouble again.

The next section on how to cure a leak will tell you what to do if your remedial action is too late, and the pipe is damaged.

Curing Leaks

A leak may occur at a compression joint because the nut has not been tightened enough. Tighten it by a series of quarter turns until the leak stops. If the reason is that the nut has been tightened too far, you will have to drain off, dismantle the joint, and fit a new olive.

A leak in a capillary joint always means draining off, breaking the joint and fitting a new one. It will have to be a compression joint unless you're prepared to wait for everything to dry out completely, as solder won't work on damp surfaces.

When you find a threaded joint leaking, you can try tightening the nut further. If this doesn't work, you will have to drain off and refit the joint.

A leak can arise in the middle of a pipe run because of frost or physical damage, such as a nail being driven into it.

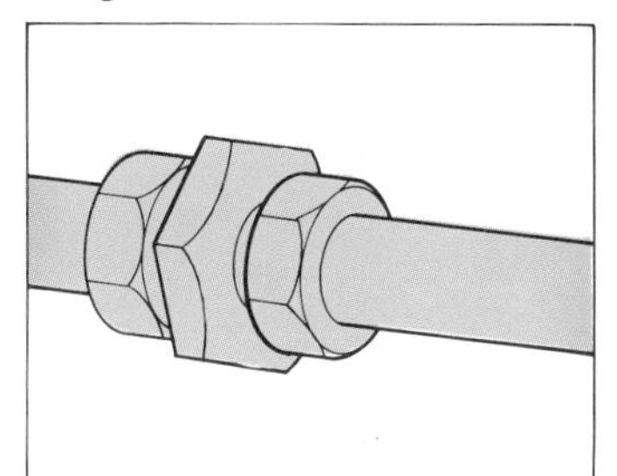

If the fracture is small, merely cut out the correct length of pipe and fit a compression joint there This is done in the same way as fitting a stop tap.

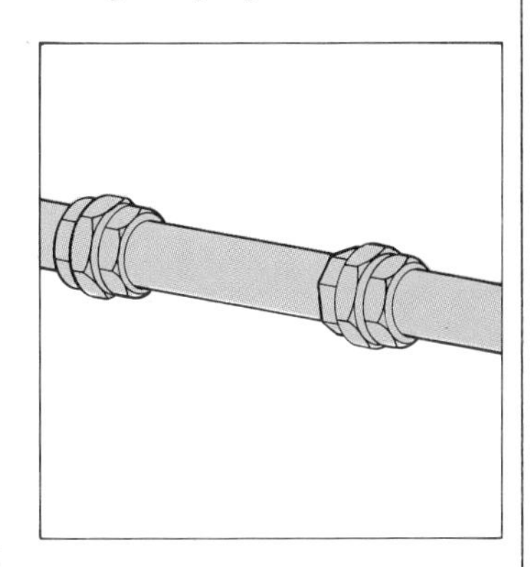

If the damaged area is too long to be covered by a joint, fit a short length of pipe by means of a compression joint at each end.

Blocked Traps

Two types of trap are found on the fittings of British homes. The *bottle trap* is found on modern plastic waste systems. It is usually held on by a knurled nut, with a rubber or plastic washer under it to ensure a watertight seal.

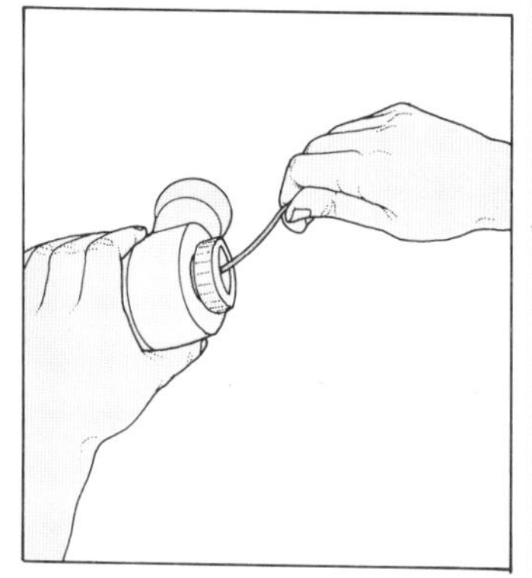

When a bottle trap becomes blocked, the easiest thing to do is to remove it and take it to another sink or basin to wash it out. You may have to poke wire through it in stubborn cases.

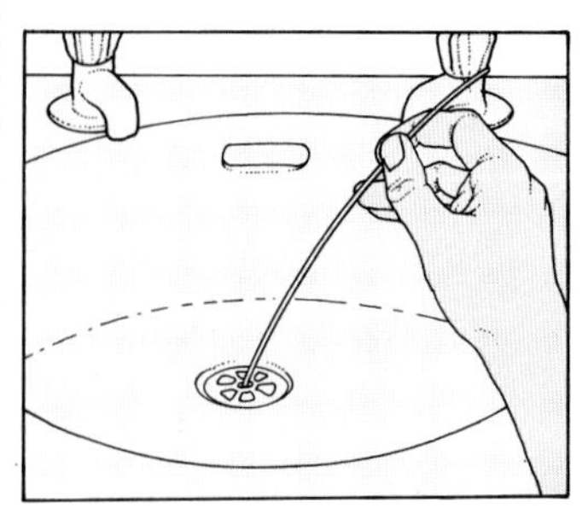

The second, older type of trap is formed simply by a bend in the pipework and is known as a *P trap* (sometimes *U trap*) because of its shape. Older types are not easily removed and it is better to get rid of the blockage from the sink end. You can try poking wire down it to clear things out, or pouring down a solution of caustic soda (be sure to follow the safety instructions on the tin).

If this isn't successful, use a plunger or rubber suction cup. In both cases, bung up the overflow with rags before you begin.

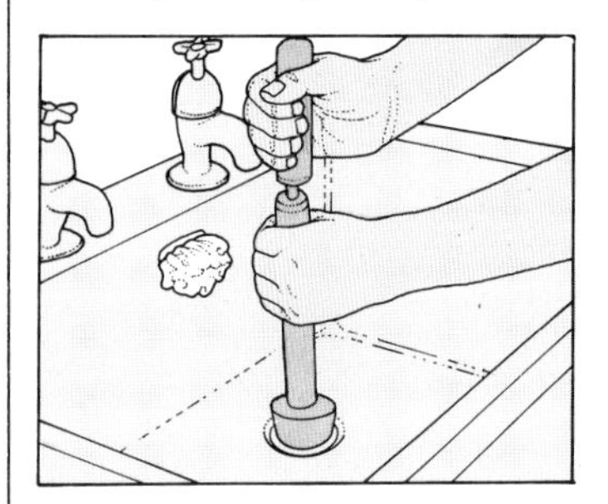

A plunger looks rather like a cycle pump. You pump the handle up and down while the other end is inserted in the waste outlet. Since water can't be compressed, it forms a column that pushes the blockage out of the way.

A suction cup operates on the same principle. It consists of a rubber or plastic flexible cup, attached to a handle. Place the cup over the outlet and move the handle up and down.

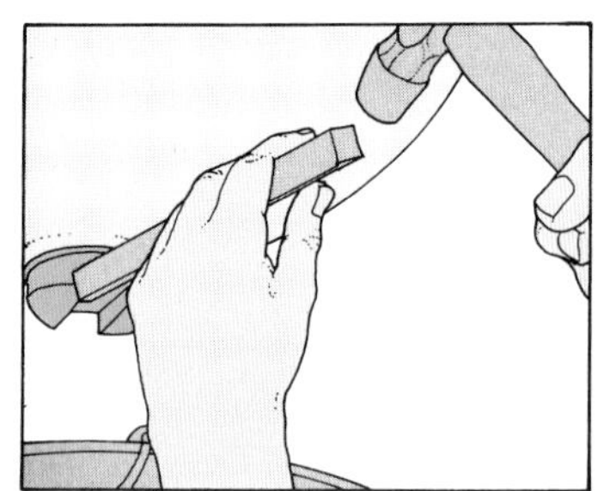

If this does not work, place a bucket under the trap and remove the cap at the bottom. The cap may take the form of a large nut that you turn with a spanner or wrench, or it may be a recessed nut with two small projecting lugs. To turn the nut, take a scrap length of wood, place one end on one of the lugs, and tap the other end with a hammer.

With the nut removed, the fall of water may bring the blockage with it. However, once again you may have to free it with a length of wire. Once the blockage is clear, replace the nut, making sure you keep the bucket underneath until you're sure it's watertight.

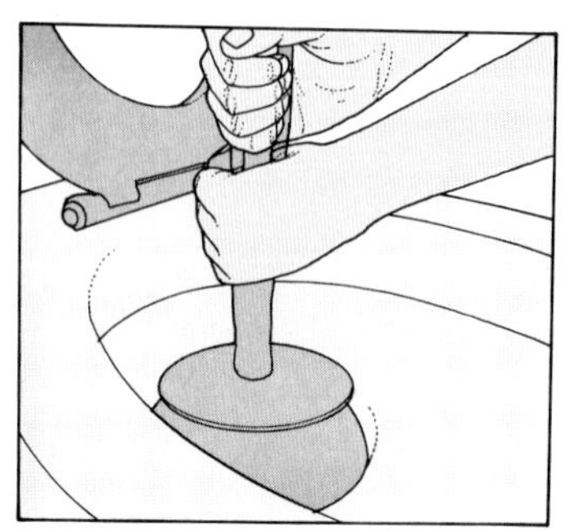

Wcs, too, have traps. A blockage can be cleared by a special plunger that has a metal plate above the rubber cup. This is to stop the cup being inverted as you work. Try to hire such a plunger. Place it in the pan and work it vigorously to force water round the bend and so clear the blockage. Once you're successful, flush the wc to re-fill the trap. If this doesn't work, call in the plumber.

Blocked Gullies

If the blockage is above the grating, scoop out as much as you can, then remove the grating and scrape it clean. Finally, wash everything in a solution of household soda, flush out with clean water and treat with disinfectant. Wear rubber or plastic gloves throughout. Treat hoppers in the same way

For blockages below ground, raise the grating and scoop out.

You may have to do this by hand, so it's even more important to wear protective gloves. Flush out clean afterwards.

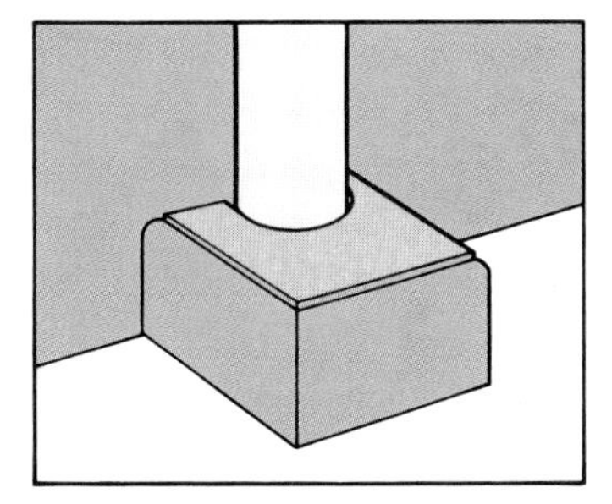

Where blockages are persistent, consider making a protective timber cover for the gulley, or perhaps having a metal one made for you.

Air Locks

If, instead of flowing freely, water splutters and hisses out of a tap, the likely cause is a bubble of air trapped somewhere in the pipe – an air lock.

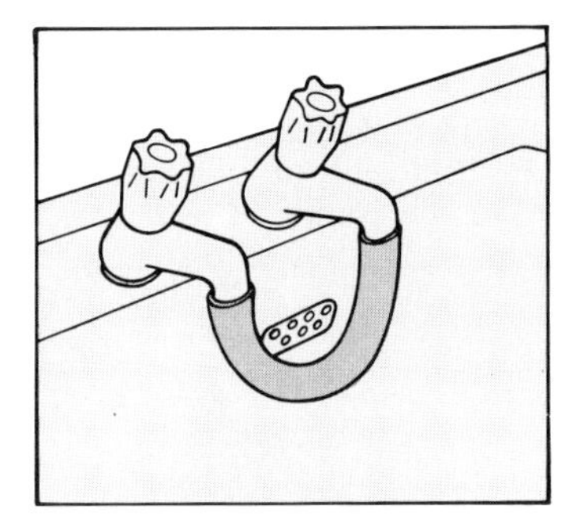

This can often be cured by running a short length of hose between the kitchen cold tap (which operates at high pressure because it is connected directly to the main) and the affected tap. This is easy enough when the affected tap is the adjacent hot tap, but it's still possible in other cases. Turn on both taps and leave for a few minutes, and then turn off – the affected tap first. The pressure from the mains should have cleared the air bubble.

Note that this dodge is frowned on by many authorities, because they say there is a danger that impure water could be sucked back into the mains. Others say that the mains pressure is so much stronger that this just won't happen. It's best to check with your local authority.

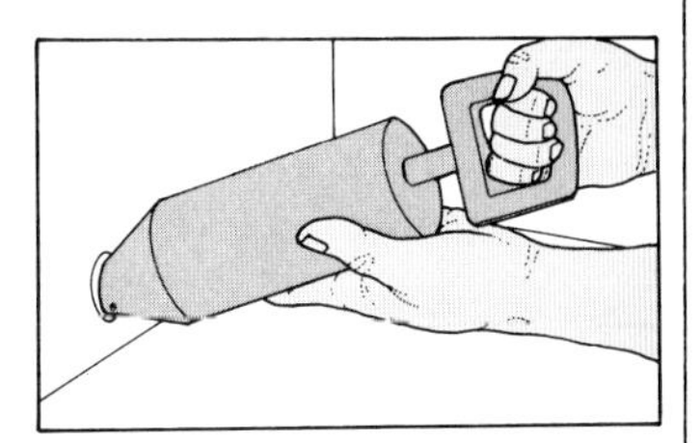

Another remedy, where the affected tap is supplied by the cold storage tank, is to use a pump. You should be able to hire one. With this, you force air down the cold tank outlet, and thus dislodge the bubble.

Radiators

Radiators are normally trouble-free, but minor faults can develop.

When a radiator feels cold, check that the boiler is working, the room thermostat is calling for heat, and that other radiators are warm. Nothing wrong there? Then the probable cause is air in the radiator. You get rid of it by opening the air vent – a process known as 'bleeding' the radiator.

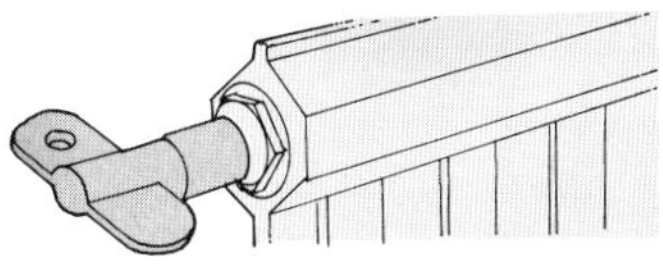

The air vent is in one side of the radiator at the top. Open it with a special radiator key – plumbers' merchants sell them. With the boiler operating, open this vent. Any air inside will hiss out. You will know all the air has been released when water starts to spurt. Keep a cup handy to catch this; it is dirty and might stain floor coverings. At the bottom of the radiator on each side is a valve – the *wheel valve*, or hand control, for turning the radiator on and off, at one side, and the *lockshield valve*, used to balance the flow of water when the system is first commissioned. The valves are supplied in two parts for installation – the radiator insert, a male fitting that is threaded into a female tapping in the radiator itself, and the main body of the valve, which is fitted to the insert. Threaded joints are used in both cases. The valve is usually connected to the pipe by a compression joint.

If a leak develops at one of the connections, try tightening the joints with a spanner – but beware of overtightening the compression joint. If this does not work, the system will have to be drained off, and the joints re-made.

Fitting a new cold tank

At one time, cold storage tanks were made of galvanised metal. But their life is limited, and eventually they have to be replaced. The replacement should be plastic or glass fibre, which won't corrode. Also, plastic ones can be squashed to go through a small loft opening.

Your local water authority will have regulations about the size of tank you must have. The size will probably be expressed as something like 50/60 gallons. The higher figure refers to the amount of water it would hold filled to the brim. The lower one is the amount it is designed to contain in practice. Usually there is a mark on the side to indicate the correct water level.

Your house will have to be without water while the work is going on, so do as much as you can beforehand.

Your first job is to take out the old tank. Turn off the rising main, and drain the tank by opening or turning on one or more of the cold taps that it feeds. Don't bother trying to free the tank by taking apart the joints connecting it to the rising main and the various supply pipes. They will probably be so corroded that you cannot turn the nuts. Anyway, it is almost certain that the pipe runs will have to be modified, no matter how slightly, to suit the new tank.

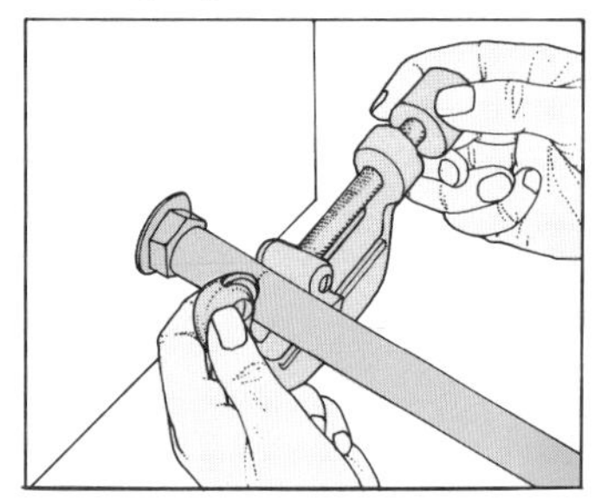

Saw through the pipes near the tank, making sure the cuts are truly square. A pipe cutter would be better for this than a hacksaw.

Be careful as you lower the old tank down from the loft. Draw-off points are positioned slightly above the base of the tank so that sediment settling at the bottom won't find its way into the supply pipes, so there will be some water left in it. The old tank may be too big to go through the loft opening; jokes about DIY plumbers who have enlarged loft openings, even removed part of a roof, to get an old tank out of the loft, are part of the plumber's folk lore. You shouldn't bother; just empty the tank and leave it up there.

Your new tank can go in the same position as the old. However, if you have a shower, remember that the pressure of water at the rose will be determined by the head of water. So, it's a good idea to consider siting the tank on a raised platform.

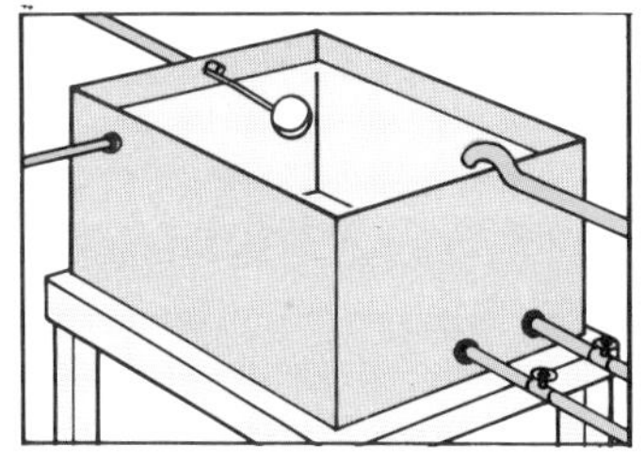

A tank full of water is very heavy, so the platform will have to be strong. Also, a sheet of thick plywood should be fixed under the flexible plastic tank.

The new tank will be supplied without holes for the pipes, since the manufacturers couldn't possibly know where these will be required. You will need at least three – one for the inlet to the ball valve, one for the overflow, and one (or more) for the outlet to supply the various cold points throughout the house. Also, the vent pipe from the hot cylinder must be positioned so that it can, if necessary, discharge into the cold tank.

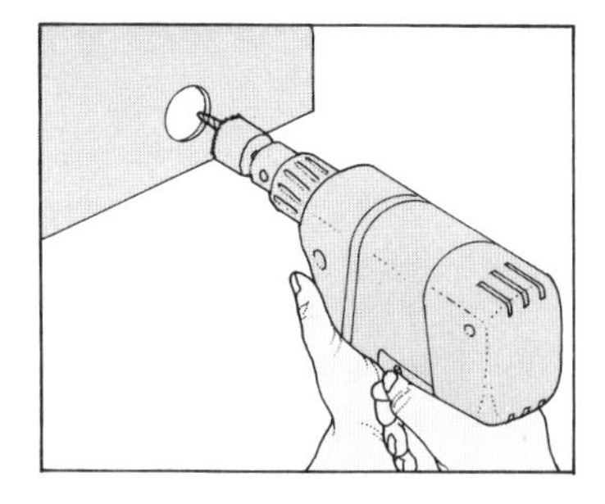

The necessary holes in the new tank should be made with a brace and bit or a hole saw in an electric drill. They should match the diameter of the pipes that have to be connected to the tank.

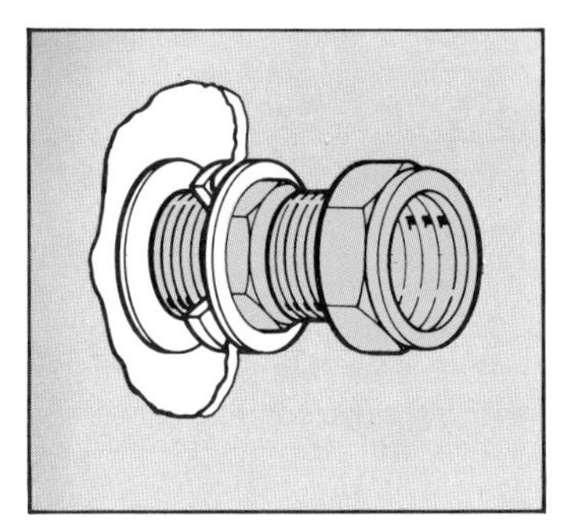

Pipes are fixed to the tank by means of *tank connectors*. These have a threaded sleeve with a back plate that goes on the inside of the tank. Push the sleeve through the hole, with a washer either side of the tank wall, and tighten the fixing nut. Pipes are connected to the outside by compression joints.

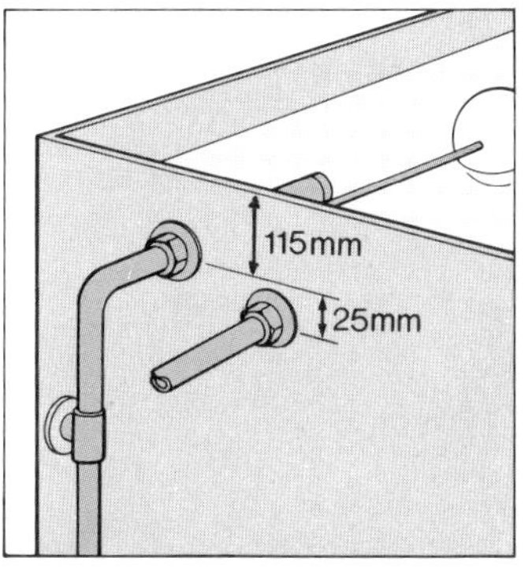

Your local water authority may have rules about the position of the holes for the pipes. However, it is normal for the ball valve inlet to be about 115mm below the rim of the tank and the overflow to be 25mm lower still. This is to prevent water from ever covering the inlet valve, with the consequent risk that it could be siphoned back into the main. The various feeds for the supply pipes should be 50mm above the bottom of the tank.

Incidentally, it is a good idea to fit stop taps to the supply pipes, so that parts of the system can be isolated when maintenance is necessary. These taps can be near the tank or, more conveniently, where you can reach them without having to climb into the loft.

The flow of water to the cold tank can often be heard throughout a small house. Cut down on the noise by closing the stopcock slightly.

Fitting an outdoor tap

An outdoor tap can be a very useful addition to the amenities of your garden, and it's easy enough to install. Remember to discuss your plans with your local water authority, who will probably want to put up your water rates.

First, choose the site for the tap – one that will be convenient in use, but not involve a complicated pipe run indoors.

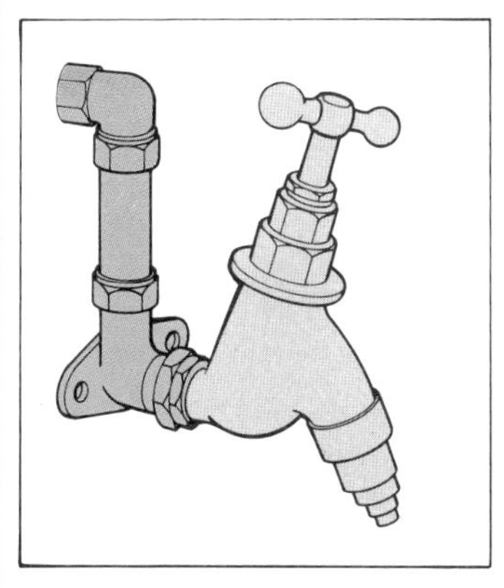

The type to use is known as a *bib tap*, which you can buy complete with hose connector. It leans away from the wall, so that you will not bang your knuckles when using it. The tap is fitted, by means of a threaded joint, to a tap connector, which is screwed to wall plugs.

Fit a 15mm T joint in the rising main, between its stop tap and the branch line to the kitchen sink. From this run a pipe to the tap connector outdoors, bending the pipe or using elbow couplings to follow the necessary route. You will certainly have to use couplings to get through the hole in the wall, for it will not be possible to push a bent pipe through it.

The hole should be slightly higher than the tap, so the pipe run outside slopes downwards. Make the hole with a club hammer and cold chisel, or you can hire an industrial electric drill and masonry bit. Work from both sides of a cavity wall, making sure dust doesn't get into the cavity.

A stop tap must be fitted in the pipe run, close to the T joint. In winter, close this tap and open the bib tap to drain off all the water from the run, so there will be no freeze-up during frosty weather.

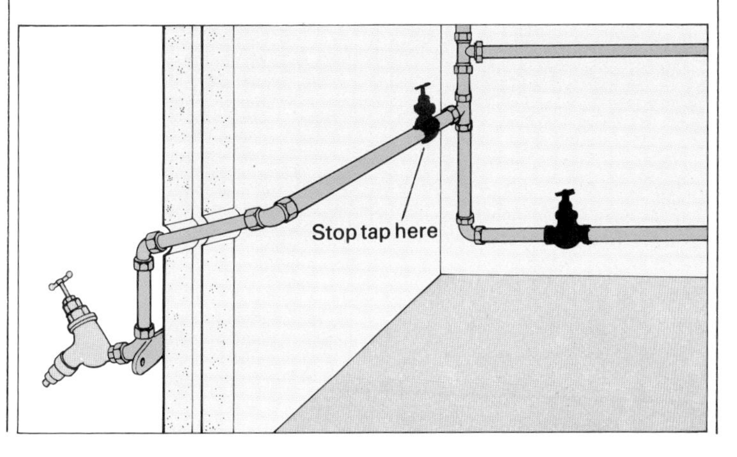

TOP TEN TIPS

1. To free a badly corroded nut, you can use heat. Depending on the situation, either wrap hot rags around the nut, pour boiling water over it, or play a blowlamp flame on it for a second or so (not plastic pipes, of course). The heat causes the metal to expand, and the movement frees the thread.

2. It's a nuisance when you want to work on a supply pipe leading from the cold storage tank, and there's no stop tap to shut it off. Well, there's no need to drain off the whole system.

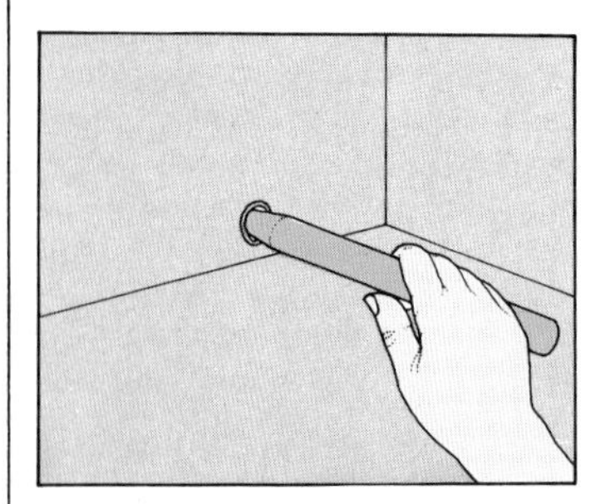

Take a short length of broomstick, sharpen it to a point at one end and push it into the outlet to the pipe inside the tank. It will act as a bung, and save you depriving the whole system of water when you want to work on just one pipe.

3. Here's how to prevent stop taps from jamming. Every six months, turn them off and on several times. Finally, open the taps fully and close them a quarter of a turn. This will restrict the flow of water slightly, but not enough to matter, and make the taps less likely to jam in future.

4. Even if your plumbing system is well lagged, frost can still strike at a waste pipe outside the house. It is no use introducing hot water into the sink or basin – heat rises and so will not go in the direction of the ice. Better to pour the hot water on the base of the pipe outside the house, so the heat will rise towards the cause of the trouble.

5. When cutting copper pipe, don't risk flattening the ends by over-tightening the vice.

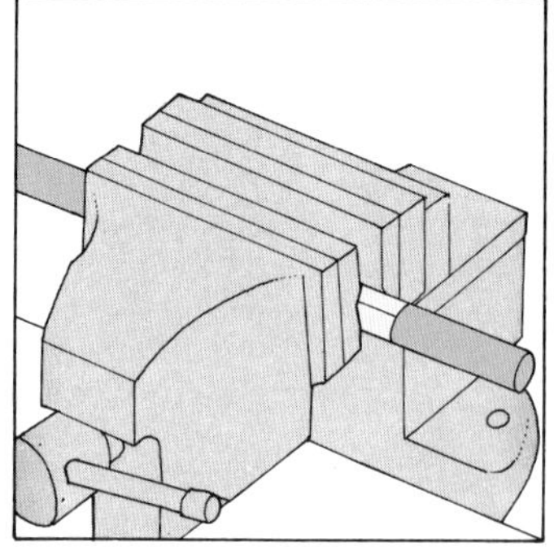

A way of ensuring an effective grip is to wrap glasspaper around the pipe before inserting it in the vice.

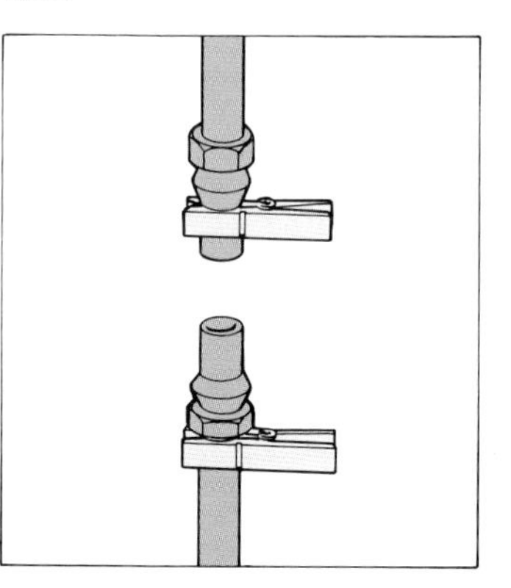

6. When fitting a compression joint in a vertical run of pipe, use spring clothes pegs to stop the nuts and olives from sliding down.

7. You may sometimes need to empty, or partly drain off, the hot storage cylinder – easy enough when the cylinder has a drain tap, but these are not always fitted. You then have to siphon the water out. Shut off the water supply, then open up all the taps to drain off the pipes.

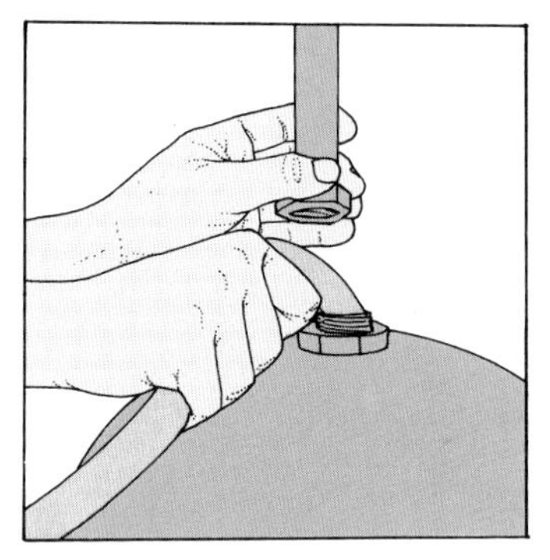

Undo the nut that connects the vent pipe to the top of the cylinder, but have a bucket and cloths ready, there will be water in this pipe.

You now need a helper. Fill a garden hose with water, pinch both ends of it so the water can't get out, and ask your helper to carry one end to a convenient discharge point. Thrust your end deep into the cylinder., Your helper can now open his end of the hose, and the water flowing down will draw the tank's water after it.

This dodge can be used for emptying many types of closed tanks. In the case of a small tank – or where a small amount of water needs to be extracted from a large one – you can use a small diameter hose.

Place one end in the tank, then suck on the other end until water starts to flow. Lower the other end, and the water will be siphoned out of the tank.

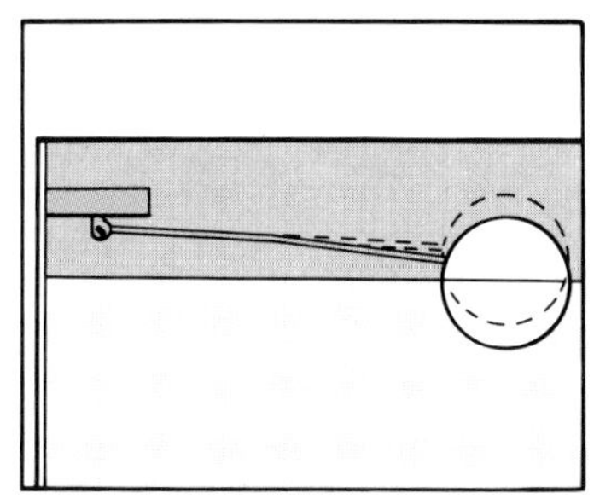

8. When a wc cistern does not give a strong enough flush, it may just be that there is not enough water in it, because the ball valve cuts off too soon. Look inside. If the water level is way below the mark, try bending the ball arm (providing it's metal) so that the ball will ride higher and admit more water.

9. Keep a quick repair kit ready to hand. A screwdriver for any grub screws in the handles, spanners for gland nuts and washers of the right size for every tap in the house.

Aim to get a tool kit that includes a spanner to fit every nut in your home's plumbing system. It is tempting to go for just one adjustable spanner to cover every situation, but these (as distinct from wrenches) are never so effective.

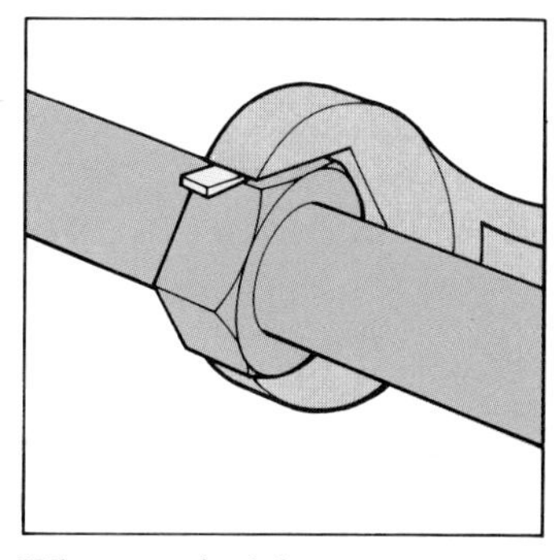

When you don't have a spanner of the right size, place a too-large spanner on the nut, then drive small pieces of wood as wedges between the jaws and the nut. They will help you to get a firm grip.

10. If a radiator valve is weeping around the spindle (not the threaded joints to the radiator or the supply pipe), there is no need to drain the system to cure it. There are two types of radiator valve (not including thermostat valves); those with O rings, and those without.

Turn the valve off, remove the handle and place cloths around the base of the pipe to catch the small amount of water that will dribble out. Remove the gland nut and turn the shaft clockwise to pull it out.

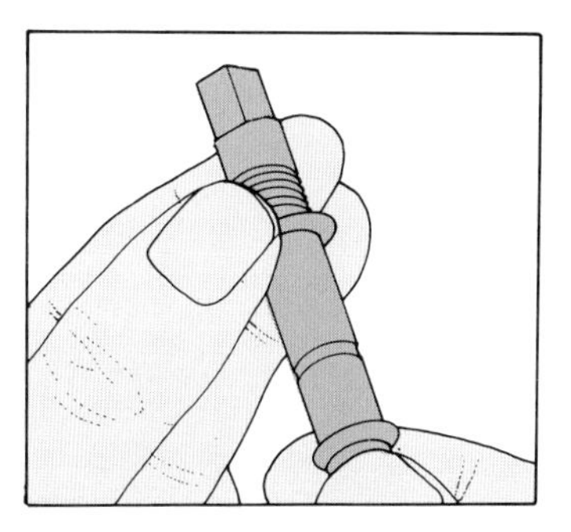

Working quickly, remove the O ring (or both if there are two) fit a new one, smear some petroleum jelly on the shaft and replace it in the valve body (twisting anti-clockwise). Replace and tighten the gland nut.

If the spindle won't come out by turning clockwise and pulling, then the valve doesn't use O rings and all you need to do is repack the body around the spindle with string and petroleum jelly. Then replace and tighten the gland nut.

Safety Tips

1. The main danger in plumbing comes when you are using a blowlamp. Take care that children and pets are well out of harm's way before lighting one.
2. Keep solder and flux out of the way of children.
3. Wear protective gloves when using the blowlamp. In particular, when using a blowlamp to form a capillary joint, make sure that the hand holding the short length of pipe is protected.
4. As well as causing burns, a blowlamp might start a fire. Bear this in mind when using one in awkward corners of the home on fixed pipework.
5. A lot of plumbing work is done in the loft. This is a very restricted space – beware of banging your head during a sudden movement. The other danger here is that of putting your foot through a ceiling. Better to stand on a short plank resting on joists than try to balance on the joists.
6. When you have to take a floorboard up to run, or work on, supply pipes underneath, always replace the board loosely when you leave the spot, even if only for a short while. Otherwise, someone may trip up over the hole.
7. Hacksaws and pipe cutters can cut skin as well as pipes. Take care when using them – and keep them away from children and pets.

Author and Series Consultant
Bob Tattersall
Editors
Dek Messecar and Alexa Stace
Design
Mike Rose and Bob Lamb
Picture Research
Liz Whiting
Illustrations
Rob Shone
Rick Blakely
Su Martin

Bob Tatersall has been a DIY journalist for over 25 years and was editor of *Homemaker* for 16 years. He now works as a freelance journalist and broadcaster. Regular contact with the main DIY manufacturers keeps him up-to-date on all new products and developments. He has written many books on various aspects of DIY and, while he is considered 'an expert', he prefers to think of himself as a do-it-yourselfer who happens to be a journalist.

Photographs from Elizabeth Whiting Photo Library, except for page 21, courtesy of Carl Boyer & Associates: Leisure Shower Cubicles, Leisure Vanity Basin

Cover photography by Carl Warner
Materials for cover photograph supplied by W. H. Newson

The *Do It! Series* was conceived, edited and designed by Elizabeth Whiting & Associates and Robert Lamb & Company for William Collins Sons and Co Ltd

First published 1984
Reprinted 1985, 1986, 1987 (twice)

Revised edition first published 1989
9 8 7 6 5 4 3 2 1

Published by William Collins Sons & Co Ltd
London · Glasgow · Sydney · Auckland
Toronto · Johannesburg

ISBN 0 00 411916 9

Printed in Spain